TREES AND FORESTS
A Colour Guide

Biology, Pathology, Propagation, Silviculture,
Surgery, Biomes, Ecology, Conservation

Edited by
BRYAN G BOWES
Formerly Senior Lecturer at the
Department of Botany
University of Glasgow, UK

MANSON
PUBLISHING

General note

All figures are supplied by the author(s) of each chapter unless otherwise acknowledged.

A CIP catalogue record for this book is available from the British Library.

For full details of all Manson Publishing titles, please write to:
Manson Publishing Ltd
73 Corringham Road
London NW11 7DL, UK

Tel: +44 (0)20 8905 5150
Fax: +44 (0)20 8201 9233
Website: www.mansonpublishing.com

Commissioning editor: Jill Northcott
Project manager: Paul Bennett
Copy editor: Janet Tosh
Layout: DiacriTech, Chennai, India
Colour reproduction: Tenon & Polert Colour Scanning Ltd, Hong Kong
Printed by: Grafos SA, Barcelona, Spain

Contents

4

Preface

A grove of sequoias [*Sequoiodendron giganteum*] – occupy an area of perhaps less than a hundred acres – The perfect specimens not burned or broken are singularly regular and symmetrical – showing infinite variety in general unity and harmony; the noble shafts with rich purplish brown fluted bark, free of limbs for one hundred and fifty feet or so, ornamented here and there with leafy rosettes; main branches of the oldest trees very large, crooked and rugged, zigzagging stiffly outward seemingly lawless, yet unexpectedly stooping at the right distance from the trunk and dissolving in dense bossy masses of branchlets, thus making a regular though greatly varied outline, – a cylinder of leafy, outbulging spray masses, terminating in a noble dome, that may be recognised while yet far off upheaved against the sky – the king of all conifers, not only in size but in sublime majesty of behaviour and port.

(John Muir, *My First Summer in the Sierra*, September 17th 1869)

A reverence and respect for trees, such as are exhibited in Muir's scientific but beautiful prose describing his experiences in the Californian Sierra Nevada is deeply embedded in the human psyche. A sacred oak (*Quercus*) was at the centre of the Greek oracle at Dodona founded in about 1800 BP; while Artemis, the goddess of woodlands, and Apollo (her twin brother) were said to be born under a date palm. For the Pehuenche Amerindians of Chile, the monkey puzzle tree (*Araucaria araucana*) is holy,

Buddha is believed to have attained enlightenment while sitting under the bodhi tree (*Ficus religiosa*), while ginkgo (*Ginkgo biloba*) is almost extinct in the wild of China but has survived as a sacred tree in temple gardens.

Sadly, for the last century or more, such respect for trees has generally been forsaken. Today, environmentally detrimental logging and forest clearances still continue, either illegally or by official connivance, driven by world population growth, industrialization, market globalization, and the pursuit of maximal (but patently unsustainable) commercial profits.

The principal aim of this present volume is to bring together and review a number of features of forests, trees, their development, natural environments, and ecology, which are usually disparately considered. These topics are expertly treated by scientists from various countries and presented in a format in which, as an integral part of each chapter, numerous high-quality colour illustrations complement the concise but clearly written text.

This Guide will be of importance and interest to anyone studying plant science, forestry, or ecology and of practical and theoretical use to conservationists, foresters, tree propagators, and tree surgeons working in the field. The 16 contributing authors, based in Britain, the USA, Australia, and Italy, reflect a wealth of expertise, which will help ensure that this volume is of value and use for students and interested amateurs in countries throughout the world.

Bryan G Bowes

Contributors

Chapters 1 and 6
BRYAN G BOWES
Formerly at the Department of Botany,
University of Glasgow, Glasgow, UK

Chapter 2
ALJOS FARJON
Royal Botanic Gardens Kew, Richmond,
Surrey, UK

Chapter 3
HUGH ANGUS
Head of Tree Collections, Westonbirt Arboretum,
near Tetbury, Gloucestershire, UK

Chapter 4
STEPHEN D HOPPER
Director, Royal Botanic Gardens Kew, Richmond,
Surrey, UK

ERIKA PIGNATTI WIKUS
Dipartimento di Biologia, Universita di Trieste,
Trieste, Italy

SANDRO PIGNATTI
Dipartimento di Biologia Vegetale,
Universita di Roma 'La Sapienza', Roma, Italy

Chapters 5 and 14
GHILLEAN T PRANCE
School of Plant Sciences, University of Reading,
Whiteknights, Reading, UK

Chapter 7
CHRISTOPHER T BRETT
Institute of Biomedical and Life Sciences,
University of Glasgow, Glasgow, UK

Chapter 8
STEPHEN WOODWARD
School of Biological Sciences,
University of Aberdeen, Aberdeen, UK

Chapter 9
CLAIRE OZANNE
School of Human and Life Sciences,
Roehampton University, London, UK

Chapter 10
PETER A THOMAS
School of Life Sciences, Keele University, Keele, UK

Chapter 11
PETER SAVILL
Oxford Forestry Institute, Department of Plant
Sciences, University of Oxford, Oxford, UK

NICK BROWN
Oxford Forestry Institute, Department of Plant
Sciences, University of Oxford, Oxford, UK

Chapter 12
DAVID THORMAN
Rostrevor, Station Road, Whittington, Oswestry,
Shropshire, UK

Chapter 13
BRENT H McCOWN
Professor, Department of Horticulture,
University of Wisconsin-Madison, Madison,
Wisconsin, USA

THOMAS BEUCHEL
McKay Nursery, Waterloo, Wisconsin, USA

Abbreviations

BP	before present
B.t.	*Bacillus thuringiensis*
B.t.k	*Bacillus thuringiensis* var. *kurstaki*
CWD	coarse woody debris
DED	Dutch elm disease
DON	dissolved organic nitrogen
E	east
HRGPs	hydroxyproline-rich glycoproteins
IBA	indolebutyric acid
IUCN	International Union for Conservation of Nature
IPM	integrated pest management
MUS	Malayan Uniform System
MVP	minimum viable population
mya	million years ago
NEP	net ecosystem productivity
NPP	net primary productivity

PGA	polygalacturonic acid
PME	pectin methyl esterase
PRPs	pathogenesis-related proteins
REGUA	Reserva Ecologica Guapi-açú
RG-I	rhamnogalacturonan-I
RG-II	rhamnogalacturonan-II
SMS	selective management system
W	west
TDF	temperate deciduous forest
TEF	temperate evergreen forest
TSS	tropical shelterwood system

Note: all sectioned specimens were photographed under the light microscope except where designated TEM (transmission electron micrograph). Abbreviations: LS, longitudinal section; RLS, radial longitudinal section; TS, transverse section; TLS, tangential longitudinal section.

Dedications

To the Editor's family

With great love to my dear, supportive, and forbearing wife Diane; in warm memory of my late wife Ruth; with all love and affection for our children Tanya and Adrian; and to my six bright, creative, and affectionate grandchildren – Sean, Aidan, Declan, Cian, Myles, and Marcus.

To William Shakespeare

The quality of mercy is not strain'd;
It droppeth as the gentle rain from heaven
Upon the place beneath. It is twice blest:
It blesseth him that gives and him that takes.
'Tis mightiest in the mightiest; it becomes
The throned monarch better than his crown;
His sceptre shows the force of temporal power,
The attribute to awe and majesty,
Wherein doth sit the dread and fear of kings;
But mercy is above this sceptred sway,
It is enthroned in the hearts of kings,
It is an attribute to God himself

(*The Merchant of Venice*)

SECTION 1 INTRODUCTION TO WOODY HABITAT

CHAPTER 1

Survey of trees: their global significance, architecture, and early evolution

Bryan G Bowes

INTRODUCTION

Most trees – apart from those growing in plantations, horticulture, arboreta, parks, or urban settings – are distributed in various biomes, or major regions, of the terrestrial world (**1**; Packham *et al.*, 1992). Each biome is characterized by its own range of animal and plant life. Chapters 2–5 of this Guide consider the various natural regions populated by trees, while Chapters 10 and 14 emphasize the vital role of trees in the complex web of life.

According to Jane (1970), there are some 20,000 species of woody plants but, more recently, Oldfield *et al.* (1998) estimated that there are possibly up to 100,000 tree (arborescent) species: these were defined as single-stemmed species growing to at least 2 m in height at maturity. However, this definition seems unduly restrictive, since it would apparently exclude some natural multi-stemmed species of

1 Overview of the Blue Mountains *Eucalyptus* forest near Sydney, Australia. This forest is dominated by eucalypts, with some 90 arborescent species of *Eucalyptus* and other members of the family Myrtaceae occurring here.

2 Multi-stemmed specimen of *Sorbus aucuparia* (mountain ash, rowan tree), which shows no evidence of a coppice origin.

3 Multi-stemmed specimen of *Eucalyptus gregsoniana* growing in Australia.

4 Foliage of the recently discovered *Wollemia nobilis* (Wollemi pine) growing in the Royal Botanic Gardens, Sydney, Australia.

5 Conical canopy of *Araucaria araucana* (monkey puzzle) bearing ripening female cones. This specimen is growing in Scotland, but the tree is endangered due to over-logging in its native southern Chile.

common trees such as *Salix* (willow), *Sorbus*, and *Eucalyptus* (**2**, **3**), as well as various palms and other taxa.

In any case, more species of tree are very likely to be discovered, or sometimes just await accurate scientific recognition and description. For example, *Wollemia nobilis* (Wollemi pine, **4**) is a relic of the extensive araucarian flora existing in the early Cretaceous period, 116 mya (million years ago). However, the Wollemi pine was only discovered in 1994; having survived until then, as a small population of conifers growing within the protective walls of a narrow, deep canyon located only 150 km from Sydney, Australia. Various new *Eucalyptus* species (Chapter 4) were only recently recognized in Western Australia.

Oldfield *et al.* (1998) concluded that, worldwide, some 8,500 tree species are classified as threatened (**4**, **5**). Of these, over 4,500 are endangered or critically endangered in their native habitats, due in

6 Specimen of the tree fern *Cyathea* sp. growing in its native temperate rain forest habitat in Queensland, Australia.

7 Tall specimen of the palm *Woodyetia bifurcata* growing in an Australian botanic garden.

large part to over-logging of natural forests, or their clearance for cultivation (Chapters 2–5 and 14). However, the continued widespread occurrence of tree cover was recently confirmed by the UN Food and Agriculture Organization, which estimated that about one third of the Earth's land surface currently remains as forest (**1**; Lamb, 2002). Nevertheless, these calculations were probably overly optimistic, since the term 'forest' was applied to tree cover of as little as 10% on any land over half a hectare in area.

TYPES OF TREES

To many European and North American observers, it is perhaps surprising that some 300 species of ferns, principally in the genera *Cyathea* and *Dicksonia*, are arborescent. Tree ferns mainly grow in tropical mountainous regions, but are also found in temperate Australasia and elsewhere. Several species of *Cyathea* may reach 20 m in height, while *Dicksonia antartica* grows up to 10 m tall. The normally unbranched trunk of the tree fern rarely exceeds 30–40 cm in diameter and bears a terminal crown of very large compound fronds (leaves), which may be several metres long (**6**).

However, almost all tree species are angiosperms, commonly termed flowering plants (**7**, **8**), or gymnosperms (**4**, **5**). The seeds of gymnosperms are naked (not enclosed within a fruit) and are usually borne in obvious cones (**9**).

By contrast, the seeds of angiosperms are enclosed

8 Tall specimen of *Betula pendula* (silver birch) growing in the wild in Scotland.

9 Large cones of the conifer *Abies procera* (noble fir) native to northwestern America but growing in Scotland.

within a fruit (**10**), which develops from the carpel (pistil) on a flower (**11**).

Gymnosperms are almost entirely arborescent. Taxonomists give varying estimates of their present-day numbers, but it seems reasonable to conclude that at least 500 conifer (**12**, **13**) and 200 cycad species (**14**, **15**) are still extant plus a single species of *Ginkgo biloba* (maidenhair tree, **16**). Of the three gnetophyte genera, only *Gnetum gnemon* is a tree. Pollination in the gymnosperms is mostly effected by wind, but beetle vectors may be involved in cycads.

The latter are tropical/sub-tropical in distribution and bear large compound fronds (**15**). These xerophytic leaves have thick, waterproof cuticular coverings and sunken pores, or stomata, concerned with gaseous exchange from the internal leaf tissue to the outside air (**17**). Cycads are generally rather squat (**15**), although *Macrozamia hopei* may attain 20 m in height (Foster and Gifford, 1974). They bear either male or female cones on separate individuals, and generally these cones are massive (**14**). The trees undergo only very limited woody (secondary)

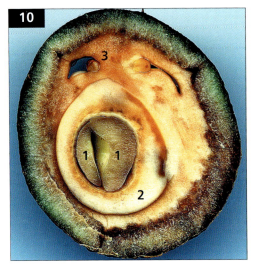

11 LS hermaphrodite flower of the arborescent dicot *Magnolia*. Note the numerous basal stamens, while abundant green carpels are inserted above on the flower axis.

10 TS of an immature fruit of the dicot tree *Aesculus hippocastanum* (horse chestnut) showing two cotyledons (1) within the ovule (2). Fruit wall (3).

12 Tall specimen of *Sequoia sempervirens* (coastal redwood and native to California) but planted in the mid-1800s in a botanic garden in Scotland.

13 Tall specimen of *Pinus sylvestris* (Scots pine) growing wild on an exposed site in Scotland. Note the long trunk (devoid of branches) but with a terminal rounded canopy.

thickening but their trunks are strengthened by the covering of persistent fibrous leaf bases (**15**).

Conifers commonly have needle-like leaves as in *Larix* (larch), *Abies* (fir), *Picea* (spruces), and *Pinus* (pines, **18**). However, some have scale-like leaves, as in species of *Cupressus* (cypress) and *Chamaecyparis lawsoniana* (Lawson cypress), or spiny leaves, as in *Araucaria araucana* (monkey puzzle). Like the cycads,

14 Massive male cone of the tropical gymnosperm *Cycas circinalis*, but growing in a botanic garden in Scotland.

15 Large specimen of the African cycad *Encephalartos altseinii* growing in an Australian botanic garden.

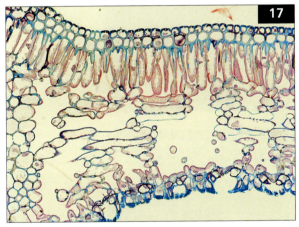

16 Newly expanding foliage of the deciduous *Ginkgo biloba* (maidenhair tree) growing in an Australian botanic garden.

17 TS of a leaflet of *Cycas revoluta* showing its thick epidermal cuticle (stained dark blue) and sunken stomata confined to the lower epidermis.

18 Foliage of *Pinus caribaea*, which is native to the Caribbean region, but is here growing in an Australian botanic garden.

conifer leaves are generally xerophytic with several modifications, such as thick, waxy cuticules and sunken stomata, to minimize water loss. Most conifers occur in the boreal, montane, and temperate rain forests of the northern hemisphere (Chapters 2 and 3). The numerous species of *Pinus* (pine, **13**) are mostly indigenous to the northern hemisphere, but *P. caribaea* (Caribbean pine, **18**) occurs in its extreme south. Conifers such as species of *Fitzroya*, *Podocarpus*, and *Araucaria* (**5**) are indigenous to the southern hemisphere (Chapter 4). With some exceptions, such as *Larix* (larch) and *Metasequoia glyptostroboides* (dawn redwood, **19, 20**), most conifers are evergreens. Their leaves are generally shed over only a few years, while the large spiny leaves of *Araucaria araucana* (monkey puzzle, **5, 21**) last up to 30 years.

Conifers often grow into very large individuals (Chapters 2 and 3) and many are extremely long-lived, as is demonstrated in the famous Mariposa Grove in the Sierra Nevada of California. Here, there are many giant trees of *Sequoiadendron giganteum*

19 A specimen of *Metasequoia glyptostroboides* (dawn redwood), a deciduous conifer native to southwest China, in its winter aspect growing in England.

20 New foliage of the deciduous conifer *Metasequoia glyptostroboides* (dawn redwood).

21 Young specimen of the conifer *Araucaria araucana* (monkey puzzle) showing its spiny, long-lived leaves on the main stem and branches.

22 Massive ancient specimen of *Sequoiadendron giganteum* (big tree, Welligtonia) growing in its native habitat at Mariposa Grove, California, USA.

(big tree, Wellingtonia, **22**). A 3,500-year-old specimen of this tree (known as 'General Sherman') is ca. 83 m tall, with a basal diameter of some 11 m. On the Californian coast, a specimen of *Sequoia sempervirens* (coastal redwood) is 111 m in height, while other North American conifers such as *Abies procera* (Noble fir), *Pseudotsuga menziesii* (Douglas fir), and *Pinus monticola* (western white pine) may be 70 m or more tall. The oldest, accurately dated living tree is a nearly 4,770-year-old specimen of *Pinus longaeva* (the bristlecone pine 'Methuselah'), growing at about 3,000 m elevation in the White Mountains of California (Anonymous, 2003). A Scottish specimen of *Taxus baccata* (yew) is estimated to be 5,000 years or more old; but it is impossible to accurately date the relics of such a large-diametered, but now much-fragmented specimen, which has none of its heartwood remaining (Lewington and Parker, 1999).

Flowering plants (trees, shrubs, and herbs) dominate most terrestrial habitats, except for the northern biomes of the moss- and lichen-rich tundra, and the conifer forests (Chapter 2). Angiosperms are divided into two main taxa (groups), the monocots and the many more numerous species of dicots, which display a number of different vegetative and floral characteristics (**23A, B**).

Monocots are mainly herbaceous and generally bear flower parts in threes or multiples of three. Their leaves are frequently long and narrow, with their main veins running lengthwise along the blade (**23A, 24**). Although most species are not woody, *Dracaena draco* (dragon tree) and some other monocots (Chapter 6) undergo an unusual form of woody secondary thickening and may form large trees. Many palms may also become tall trees (**7**) and some species live for more than a century (Tomlinson, 2003). Palms do not undergo secondary thickening, but instead increase in thickness by a process of diffuse secondary growth throughout the living tissues of the trunk (Esau, 1965).

By contrast, in dicots the flower parts usually occur in fives or fours, or are indefinite in number (**11, 23B**) and most species show some degree of

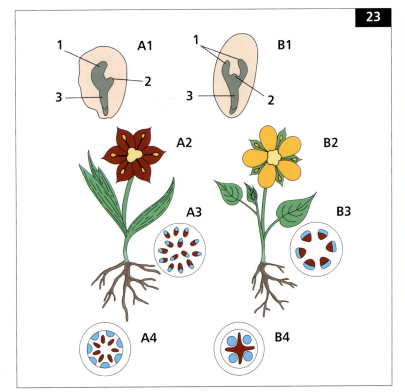

24 Leaf of a large-leaved bamboo showing the arrangement of parallel veins in its leaf, which is typical of the venation in a monocot.

23A–B Idealized diagrams illustrating the salient differences between mono- (series **A**) and dicotyledonous (series **B**) flowering plants. **A1/B1**: LS seeds; cotyledon (1), plumule (2), radicle (3). **A2/B2**: Leaf and floral differences.
A3–4/B3–4: distribution of the primary xylem (red) and phloem (blue) in the young shoots (**A3/B3**) and roots (**A4/B4**).

25 Leaves of *Firmiana malayana* (family Sterculiacea); although a tropical dicot tree, it is deciduous, shedding its leaves after a dry period and remaining bare for up to three months.

26 Leaf of the dicot *Acer pseudoplatanus* (sycamore) showing its branched veins which terminate in a reticulum of fine veinlets. Note also the numerous insect galls.

27 Newly expanding foliage from a breaking bud of *Aesculus hippocastanum* (horse chestnut) in spring.

28 Stump of the conifer *Pinus sylvestris* (Scots pine) clearly showing its annual rings.

woody thickening (Chapter 6). The leaves are generally relatively wide, and hence a dicot tree is commonly termed broadleaved (25). From the main vein(s) of the leaf, prominent side veins arise and further branch with the ultimate veinlets, forming a closed network in the leaf blade (26). On average, some 25–30 different broadleaved species occur per hectare of temperate forest, whereas the number of tropical tree species is much greater, with up to 300 present per hectare (Thomas, 2000; see also Chapter 5). Pollination of broadleaved trees is primarily effected by insects and, to a lesser extent, by birds and mammals, but wind pollination is common in species native to higher latitudes.

Temperate broadleaved trees are usually deciduous (Chapter 3), producing new leaves when the buds burst in spring (27). The foliage senesces and drops in autumn (the fall), with the accompanying dormancy of the tree. In seasonally dry tropical climates, trees such as *Firmiana malayana* (25) may shed their leaves during the dry season. However, most tropical species are evergreen (Chapter 5), and the woody thickening of their trunks and branches usually proceeds without abrupt interruption (Longman and Jenik, 1987). By contrast, in temperate broadleaved species, and also most coniferous trees, the annual growth rings mark the cessation of woody thickening at the end of the growing season in late summer/early autumn and its resumption in the following spring (28; see also Chapter 6).

IMPORTANCE OF TREES FOR CIVILIZATION

Trees and timbers have been used all over the world, both prehistorically and in historical times, in the construction of shelter and housing (**29, 30**). Today timber – together with a large variety of wood products – is used worldwide, both for domestic housing and for larger-scale constructions. Wooden-framed buildings (**31**) house more than 70% of the populations of Australia, Japan, North America, and Scandinavia. Such construction is much less popular in the UK but, even here, in a typical brick-built modern family house, wood is used for the rafters, trusses, joists, stairs, doors (and often the window frames), while the floors are either of timber or wood composites (Lewington, 2003). In developing countries of the tropics, smallholder farmers utilize agroforestry and cultivate various species of trees, as well as harvesting plants from the wild. These provide fruit, animal fodder, timber, poles, firewood, livestock fences, and shade, all in combination with the nurture of their livestock and food crops (Longman and Jenik, 1987).

In Burkina Faso (West Africa) the tree *Bombax costatum* (locally known as li-suoli) provides a profusion of edible or other useful products for the villagers. The flowers attract bees, which provide honey, the dried outer parts of its flowers make a mineral-rich sauce, the fruit is used in brewing, the floss around the seeds gives kapok; the bark yields a brown dye, the leaves provide animal fodder; and the timber is used for the fishermen's canoes. Among other species harvested from local trees for food are the leaves of *Adansonia digitata*, *Balanites aegyptica*, and *Commiphora africana*, and fruits from *Tamarindus indica* and *Saba comorensis* (Anonymous, 2005).

Many species of both conifers and broadleaved trees yield important commercial timbers. Conifer woods lack thick-walled hard fibres and are classed as softwoods. This term is somewhat misleading, since the water-conducting cells (tracheids) in the wood are often thick-walled and tough (Chapter 6). *Taxus baccata* (common yew), for instance, possesses a very strong, durable, and even-textured wood. Yew branches were used extensively in the Middle Ages for the production of the English longbow, while the hard wood is excellent for use in

29 Reconstruction of wooden crannoch (communal dwelling) built for the protection of its inhabitants over the margins of a loch, as used in the late Bronze/early Iron Age of Scotland.

30 The interior of the Great Hall of Stokesay Castle in England, which still retains some of the original massive oak trusses holding up its roof.

31 Exterior of a modern wooden-framed bungalow in Queensland, Australia (courtesy of Molly Miller). Despite the danger from termites, wood is still a very popular building material for Australian housing.

turnery. Broadleaved trees usually contain a large proportion of tough, thick-walled wood and fibres (Chapter 6), and hence are termed hardwoods. However, the soft, light, and easily carved wood of the tropical broadleaved *Ochroma pyramidale* (balsa wood) provides an obvious exception.

In the boreal (taiga) forests of the northern hemisphere across the far northwest of Europe, Russia, and North America, huge areas of natural conifer forests still exist (Chapter 2) but many have been very heavily logged. In Scandinavia and Finland most of the original conifer trees are now replaced by commercially managed secondary forests, and in northern Europe large areas of conifers such as *Picea abies* (Norway spruce), *P. sitchensis* (Sitka spruce), and *Pseudotsuga menziesii* (Douglas fir) have been planted. In the northwestern USA, plantations of *Abies* (firs), *Picea glauca* (white spruce), *Pinus* (pines), *Tsuga heterophylla* (western hemlock), *Pseudotsuga menziesii* (Douglas fir), and other conifers are widely established and provide very valuable commercial timbers. In the southern hemisphere various conifers, including *Araucaria araucana* (5), *Fitzroya cupressoides*, and species of *Agathis* and *Podocarpus*, have been logged for timber, sometimes to near extinction in the wild (Chapter 3). In addition to timber for construction,

a vast quantity of conifer wood pulp, mainly from *Picea* (spruces), is processed into newsprint, cardboard, fibreboard, and rayon.

Several broadleaf trees provide important temperate and tropical hardwood timbers (Chapters 3–5). Some may attain great heights and *Eucalyptus diversicolor* (Karri), formerly an extensively logged native of southwest Australia but now protected, grows up to 87 m tall and 4 m in trunk diameter. In Tasmania, a 350-year-old specimen of *E. regnans* (called 'El Grande') was discovered in 2002. It was 79 m tall, with a girth of 20 m and an estimated volume of some 439 cubic metres (Anonymous, 2004). The tree was accidentally fire-killed (but still remains standing) by a mismanaged forestry regeneration burn, which spread from an adjacent logged area. In Tasmania this magnificent tree species is still being cut down to provide woodchip for the paper industries in Japan and elsewhere.

There are numerous other important hardwood trees, with some further examples being various species of *Carya* (hickories), *Dryobalanops* (kapur, camphor wood), *Fagus* (beech), *Fraxinus* (ash), *Grevillea* (silky oak), *Juglans* (walnut), *Khaya* (African mahogany), *Liquidambar* (satin walnut), *Magnolia* (magnolia), *Populus* (cotton wood, poplar), *Quercus* (oak), *Swietenia* (American mahogany), *Tectona grandis* (Asian teak), *Tilia* (lime, linden), and *Ulmus* (elm). Such trees provide very valuable timber and are extensively used for buildings/joinery, furniture, and sculpturing (30, 32). Smaller trees and branches are also harvested to provide poles, fuel, and charcoal (33). Finer branches from species such as *Betula* (birch), *Corylus* (hazel), *Salix* (willow), and other genera, often from trees which have been pollarded or coppiced (34, 35), are used for brooms, baskets, and fencing (36).

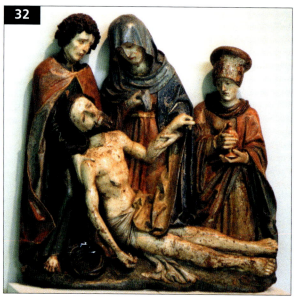

32 16th century carved limewood (*Tilia* sp.) altar piece from southern Germany.

33 Demonstration of a charcoal kiln in a national park in southern Spain.

34 Ancient pollard of the broadleaved *Salix* sp. (willow) growing along the river Cam in England.

35 Young coppice branches growing from a stool of the broadleaved tree *Acer pseudoplatanus* (sycamore).

36 Demonstration of a wicker fence, woven from coppice branches of *Corylus avellana*, at a crannoch reconstruction in Scotland (see also 29).

Many broadleaved trees provide valuable foods and spices for human consumption (37–42). Some important examples from temperate and tropical regions are *Artocarpus* spp. (breadfruit and jackfruit), *Bertholletia excelsa* (Brazil nut), *Carica* spp. (papaws, mountain papaw), *Castanea* spp. (sweet chestnut, Chinese chestnut), *Cinnamomum verum* (cinnamon), *Coffea* spp. (Arabica, robusta, and liberica coffees), *Citrus* spp. (grapefruit, lemon, lime, tangerine, and orange), *Ficus carica* (edible fig), *Garcinia mangostana* (mangosteen), *Malus* ×
domestica (apples, in a large variety), *Mangifera indica* (mango), *Nephelium lapaceum* (rambutan), *Olea europea* (olive), *Pistacia vera* (pistachio), *Prunus* spp. (apricot, cherry, damson, nectarine, peach, and plum), *Syzgium aromaticum* (clove), and *Theobroma cacao* (cocoa).

In addition to such broadleaved trees, the monocot palm family, which is mainly confined to the tropics and sub-tropics, contains a number of trees which provide very important foods. Coconuts are gathered from *Cocos nucifera* (43), dates from

37 Crop of *Citrus aurantium* (Seville orange) growing on a southern Spanish street.

38 Tropical breadfruit on the small broadleaved tree *Artocarpus*.

39 Apple tree (*Malus* x *domestica*) bearing ripe fruit in an English autumn.

40 Fruits of *Juglans nigra* (black walnut) – a broadleaved tree native to North America but here growing in England.

Phoenix dactylifera (**44**), and sago from the trunk of the *Metroxylon sagu*. *Metroxylon rumphii* and the fruits of *Elaeis guineensis* yield both palm oil and palm kernel oil. Various palm species are also tapped for their sugary sap, which may be used as a sweetener or fermented into toddy. Palm hearts (terminal buds) are harvested from various palms, and provide a vegetable delicacy. Many palms are also of importance for housing, with their trunks used for beams and flooring, and the leaves for roofing and weaving. Species of *Pandanus* (screw pines, another tropical family) yield large fruits (**45**), which are roasted and their seeds eaten by the Australian Aborigines. The long fibrous leaves of *P. utilis* are also used in thatching and to make Manila

41 Fruits of the tropical broadleaved evergreen tree *Garcinia mangostana* (mangosteen) – some fruits have been opened to show their white arils containing the edible seeds.

42 Fruits of the tropical broadleaved tree *Nephelium lapaceum* (rambutan) with some fruits opened to show their edible arils.

43 Crop of coconuts borne among the terminal crown of palm leaves on the arborescent monocot *Cocos nucifera*.

44 Crop of edible date fruits from the tropical palm *Phoenix dactylifera*.

45 Fruit of the arborescent monocot *Pandanus* sp., which provides a food sometimes eaten by the native (aboriginal) Australians.

46 Edible seeds and fan-shaped foliage of the gymnosperm *Ginkgo biloba* (maidenhair tree).

47 Mature seeds of the shrubby cycad *Cycas circinalis*. The seeds are highly toxic (as are all cycad seeds) but after prolonged leaching in water they can be safely eaten.

hats. The long, hollow, and very strong stems of several tall tree-like bamboos, such as *Dendrocalamus* spp., are used extensively to build houses and bridges, for scaffolding in building construction, and also as water pipes. The newly emerged shoots of bamboos, such as species of *Bambusa*, *Dendrocalamus*, and *Phyllostachys*, are harvested when about 15 cm long and, after removal of their enclosing leaf sheaths, are widely cooked as a vegetable in Asia.

The seeds of a number of conifers and some other gymnosperms are edible. Examples include various species of *Pinus*, *Araucaria* (bunya-bunya pine, monkey puzzle, and parana pine), *Gnetum gnemon* (gnetum), and *Ginkgo biloba* (ginkgo, **46**). In some Pacific Island countries, and also Australia, the indigenous people gather starchy seeds of *Cycas* (**47**) and other cycads such as *Lepidozamia peoffsykana*. After prolonged leaching in water, the seeds can be ground into flour. However, the fresh untreated seeds are very poisonous for humans (Chapter 4).

Several plant commodities are derived from the bark of trees; such as cork from *Quercus suber* (**48**),

fabric and rope from *Adansonia* spp., and rubber from *Hevea brasiliensis*. Various drugs and medicines are also sourced from trees. Quinine is derived from the bark of *Cinchona officinalis*, while an extract from the bark of the *Prunus africana* is used to treat prostatic disorders. The latter species is slow-growing and classified as endangered due to over-felling in the wild. In Kenya, a single tree may fetch around £100 for its bark (Lewington, 2003). Many other trees yield a variety of medicines. Aspirin was originally isolated from *Salix alba* (white willow), while various antiseptics are also derived from trees, for example tincture of witch hazel from *Hamamelis virginiana*, tea tree oil from *Melaleuca alternifolia*, and eucalyptus oil from species of *Eucalyptus*. Recently, the drug Taxol has been isolated from *Taxus brevifolia* and *T. baccata* (Pacific yew and common yew, **49**) and is used in the treatment of breast and ovarian cancers.

Trees and forests are very important as amenities, particularly in the developed world, for activities such as walking, hunting, riding, picnicking, and general relaxation. In the past, and still to some

48 Stacked cork from the bark of the evergreen broadleaved tree *Quercus suber* (cork oak), after harvesting in Spain.

49 Foliage and 'berries' of *Taxus baccata* (common yew). The foliage and red 'berries' are sometimes eaten by cattle with fatal results but the foliage contains a potent anticancer compound which, when extracted, is marketed as Taxol.

50 Specimen of a large evergreen broadleaved tree *Ficus* sp. (fig) – this is regarded by Buddhists as a sacred tree, as indicated by the ribbons tied around its trunk and small figures of Buddha embedded in its roots.

extent today, trees have had deep spiritual/religious significance for humans. In Europe, oaks (*Quercus robur* and *Q. petraea*) were important in Germanic and Norse mythology, while in Britain the Celts believed oak trees to be sacred. Also, throughout Europe, the evergreen, long-lived, and highly poisonous common yew (*Taxus baccata*, **49**) must have been of great significance to our ancestors. Ancient specimens grow in many churchyards, and it is likely that some of these churches were established on what were originally pagan religious sites (Lewington and Parker, 1999). The relic yew, still just surviving in the church graveyard at Fortingall in Scotland, greatly pre-dates the coming of Christianity to Scotland. In Asia, several species of fig, in particular *Ficus benghalensis* (banyan tree) and *F. religiosa* (bodhi tree), are revered. Buddha is believed to have gained enlightenment while sitting under a specimen of the latter tree. Such figs still often have images of Buddha embedded in their spreading roots, while the trunk is bound round with yellow and red ribbons (**50**). *Ginkgo biloba* (ginkgo, **46**) still exists in the wild, but as an endangered species, in a mountainous area of Zhejiang, China. However, it has also survived in Chinese, Korean, and Japanese temple gardens, having been planted there and tended by Buddhist monks. In Chile, *Araucaria araucana* (monkey puzzle, **5**) is a sacred tree for the Pehuenche Aborigines, with its nutritious seeds forming a staple of their diet.

TREE ARCHITECTURE

TRUNK AND CANOPY OF THE AERIAL TREE

In most trees it is estimated that the trunk (**51**) makes up 40–60% of their mass, while the aerial canopy (which bears the foliage, cones, flowers, and fruit; **5, 8, 25, 37**) contributes another 20–30% (Thomas, 2000). Frequently, the height of a tree is approximately 100 times its trunk diameter, but there are various exceptions to this rule of thumb (**52**). Many conifers retain a single main trunk, from which arise regular rings of branches (**53**). The mature conifer tree canopy often retains a regular pyramidal form (**12, 19**), due to its continued growth from a dominant terminal bud. Boreal and montane conifers (Chapter 2) often have branches that slope downwards and from which the snow tends to slide off, while some species of *Abies* (fir) and *Picea* (spruce) have broad canopies when growing in deep shade or at treeline level (Thomas, 2000). Tree architecture often changes with age. For example, *Pinus sylvestris* (Scots pine) is more or less conic when young but, particularly in exposed sites, many of its lower branches have been lost by the time it attains maximum height. Its long trunk is often then left bare, except for its terminal canopy (**13**).

Palms and some other arborescent mono-cotyledons frequently show an unbranched trunk bearing a terminal tuft of large, elongated leaves (**7, 43**). The natural form of broadleaved trees is very variable, and their canopies may range from somewhat spherical to broadly spreading (**1**). The latter shape is commonly seen in European temperate species (**8**). In marked contrast, some broadleaved trees are narrowly or broadly columnar, with examples seen in *Fagus sylvatica* 'Fastigiata' (Dawyck beech, **54**), *Populus nigra* 'Italica' (Lombardy poplar), *Nothofagus cunninghamii* (myrtle beech), and *Chrysolepis chrysophylla* (golden chestnut). *Polyalthia longifolia* (Indian mast tree) has a weeping habit. Various 'architectural models' have been proposed to encompass the extensive range of tree forms, but it seems that each species conforms to one of 25 possible characteristic branching patterns (Bell, 1991).

According to Thomas (2000), the shape of a tree is also a compromise between various factors, such as the overall exposure of the foliage to optimize photosynthesis, reproduction (the display of flowers for pollination and subsequent seed dispersal), the most 'cost-effective investment' by the tree in growing its large wooden skeleton, and the response of the tree to its immediate physical surroundings. Hence, in addition to its basic 'architectural model', the shape of a tree is often greatly modified by external environmental factors, such as grazing and wind (**13, 55–57**), disease (**58**; see also Chapters 8 and 9), and pruning/tree surgery (**34, 35, 51**; see also Chapter 12).

51 Massive ancient pollarded specimen of the broadleaved deciduous tree *Quercus robur* (English oak) growing in the wood pasture of Hatfield Forest, England.

52 Huge swollen trunk of the broadleaved sub-tropical tree *Brachychiton rupestris* growing in an Australian botanic garden.

53 Trunk of *Pinus sylvestris* (Scots pine) showing how several side branches all originate in whorls at the same level from the trunk.

54 Specimen of the broadleaved *Fagus sylvatica* 'Fastigiata', which in 1860 originated as a sport in the woodland at Dawyck, Scotland.

55 Group of wind-trimmed conifers ('Krumholz') exposed to fierce gales on the tree line at Libby Flats, Wyoming, USA.

56 Grazed and wind-trimmed miniature specimen of *Fraxinus excelsior* (ash) growing in a gryke of the limestone pavement on the Burren Region of western Ireland.

57 Major branch breakage on a large specimen of *Quercus petraea* (sessile oak) growing in native mixed woodland in Scotland.

58 Canopy of *Pinus sylvestris* (Scots pine) with a huge foliage gall probably caused by infection with the fungus *Taphrina*.

UNDERGROUND ROOT SYSTEM OF THE TREE

Tree roots generally lie underground and hidden from view, unless exposed by erosion in a wind-blown tree or on a very shallow soil over bedrock (**59, 60**). Nevertheless, the complexly branched root system is a vital component of the tree. Water and nutrients are absorbed from the soil via the roots, and anchorage for the aerial trunk and canopy is also provided by the root system. In thickened roots, an outer layer of impermeable cork is present, but the lenticels are channels through which gaseous exchange with the soil atmosphere can occur (**61**). In many trees, the root system is estimated to make up 20–30% of the tree mass, but the figure may be as little as 15% in some tropical trees and up to 50% in trees growing in very dry conditions (Thomas, 2000).

A tree root system generally shows a central plate of framework roots (**60**), which then spread laterally as far as the margins of the aerial leaf canopy, where they are much narrower and are primarily concerned with the absorption of water and its dissolved minerals. At their origin from the trunk, the framework roots may be 30 cm or so thick, and frequently graft together where they cross each other, so producing a rigid root plate which binds the associated soil and stones together (**59, 60**). The swollen root bases often produce a prominent collar at the base of the trunk which, even in temperate trees, may flare into small buttresses (**62**). However, in many tropical trees, this is much more marked and the root buttresses may extend several metres up the trunk and slope for some distance obliquely into the soil (**63**).

According to Rackham (2004), most trees in England are shallow-rooted, with the tap root not persisting after early seedling growth. This is also frequently likely to be the case in Europe and elsewhere, with such trees very susceptible to gales. In the woodlands of western Scotland many wind-thrown, prostrate, but living trees still survive after gales from some decades earlier. These continue to grow actively but with only very limited root contact with the soil (**64**). A similar situation occurs when a standing tree is root pruned (as in site clearance for construction or road building). It is estimated that in many cases a tree may survive when only a quarter (sometimes less) of its root system still remains (Rackham, 2004). However, although such a restricted area of roots may be sufficient for the water requirements of the tree, it will often be insufficient for its role as a stabilizing support for the aerial tree. In areas with only a thin soil cover, the root plate lies superficially (**59**), and the expansion of these roots that accompanies secondary thickening helps to fragment and weather the surface of the underlying bedrock (**65**). In deeper soils the root plate may be situated a metre or more beneath the surface, while sinker roots may further grow down to a deeper lying water table.

59 Underside of the root plate of a wind-felled specimen of *Fraxinus excelsior* (ash) growing in mixed native woodland in Scotland.

60 The root plate of *Quercus petraea* (sessile oak) felled by wind-blow, and then exposed after flooding by the waters of Loch Lomond, Scotland.

61 Detail of a main root of *Quercus petraea* (sessile oak), exposed after flooding/soil erosion, showing numerous lenticels in its bark.

62 Young specimen of the deciduous broadleaved tree *Fagus sylvatica* (beech), demonstrating that even temperate trees may develop root buttresses.

63 Massive root buttresses on the tropical broadleaved tree *Parkia javanica*.

64 Large wind-blown specimen of *Quercus petraea* (sessile oak) in Scotland, which was probably felled by the devastating gales of 1987/1990, now showing the transformation of a side branch into a large vertical new trunk despite its meagre root connection to the soil via the original root plate.

65 Part of the exposed root system of *Fraxinus excelsior* (ash) growing in rock fissures and helping to break up the bedrock.

ACCESSORY ROOTS ON THE AERIAL TREE

Some trees develop trunk- or branch-borne adventitious roots. As well as providing an increased aerating and absorbing surface (**66, 67**), these may serve as an additional anchorage/support to the trunk, as in species of *Ficus* and *Pandanus* (**68, 69**), and also facilitate the spread of the tree following branch rooting (**70**).

Tidal mangroves have various anchorage and aeration devices, such as the complex stilt (prop) roots of *Rhizophora* spp., the upright peg roots of *Avicennia* spp., and the knee roots of *Bruguieria* spp. (**67**). At low tide, atmospheric oxygen diffuses through the lenticels into the well-developed cortical intercellular space system of the subterranean roots. Knee roots also develop in the gymnosperm *Taxodium distichum* (swamp cypress, **66**).

Pillar and strangler root systems occur in a number of species of fig (*Ficus*, **69, 71**). A specimen of *F. benghalensis* (banyan tree), planted in 1782 in Calcutta, now forms a tree grove comprised of some 1800 separate pillars (trunks), which originated as branch-borne adventitious roots (Thomas, 2000). The roots of the strangler figs originate from a seed(s) germinating in the canopy of a host tree. These roots repeatedly branch and anastomose as they grow down the trunk and eventually reach the soil. The host trunk often becomes completely enclosed ('strangled'), dies, and then rots away. The shell of fused *Ficus* spp. roots remains (**71**) and these thicken to form an independent hollow-trunked fig tree.

Many old trees are hollow, and very commonly the growth of internal adventitious roots within the trunk allows recycling of nutrients released from the rotting heartwood. In the UK this is especially common in ancient pollards of tree such as *Fagus sylvatica* (beech, **72**), *Fraxinus excelsior* (ash), and *Carpinus betulus* (hornbeam). In other trees, the rooting of low branches (layering) may eventually

66 Knee roots of the conifer *Taxodium distichum* native to swampy areas of the southern USA.

67 Breathing roots in tropical broadleaved trees. Knee roots of *Bruguieria gymnorhiza* and upright air roots of *Avicennia marina* exposed at low tide in the coastal mud flats of Queensland, Australia.

68 Prop roots of the arborescent monocot *Pandanus pedunculatus* growing along the sandy coast of New South Wales, Australia.

lead to the propagation of a new individual (70). Root suckers (shoots forming on roots) occur on many species of broadleaved tree, such as *Ilex* (holly), *Ulmus* (elm), and *Robinia pseudoacacia* (false acacia). Clonal stands of thousands of poplar and aspen (*Populus*) have originated in this manner in North America. Some conifers, such as species of *Abies* (fir), *Picea* (spruce), and *Juniperus* (juniper), also propagate by root suckers, and *Sequoia sempervirens* (coastal redwood, 73) also propagates vegetatively in this way.

69 Specimen of the evergreen broadleaved tree *Ficus benghalensis* (fig) growing in an Australian botanic garden, showing its numerous pillar roots.

70 Large layered branch of *Fagus sylvatica* 'Purpurea'. The branch has rooted and is considerably thickened compared to its diameter (arrow) near its origin from the tree trunk.

71 The fused roots of a *Ficus* (strangler fig) in Queensland, Australia. Note that the trunk of the original 'strangled' host tree has rotted away.

72 Top of a hollow trunk of an ancient pollarded specimen of *Fagus sylvatica* (beech), showing large adventitious roots growing down its hollow core to tap nutrients released from the rotted heartwood.

73 Adventitious shoots sprouting from roots of the conifer *Sequoia sempervirens* (coastal redwood).

THE ROOT SYSTEM AND
MICRO-ORGANISM SYMBIONTS

In many plants, the roots develop fine hairs just behind their growing tips (**74**). These greatly increase the absorptive surface of the root. However, root hairs are frequently sparse in trees and may be absent from some gymnosperms. Nevertheless, most trees develop accessory root/micro-organism symbiotic systems of nutrient supply and absorption. These take the form either of root nodules (**75–77**) or of endo- and ectomycorrhizae (**78–80**). In *Alnus rubra* (red alder) and some legumes, nodules and mycorrhizae can occur on the same root system. Some trees can form either endo- or ectomycorrhizae and, occasionally, as in *Eucalyptus*, *Juniperus* (juniper), and *Salix* (willow), both types may be present on the same tree.

Root nodules are inhabited by symbiotic micro-organisms (**77**), which fix or transform atmospheric nitrogen in the soil into nitrogenous compounds utilizable by the tree. In leguminous trees, such as species of *Acacia* and *Laburnum*, and also *Ulmus* (elm) and other members of the elm family, the bacterium *Rhizobium* occurs in the nodules. In *Alnus* (belonging to the birch family, in which other nodulated species also occur) the nodules may live up to 10 years and grow large (**75**), sometimes to tennis ball size. Nodules in *Alnus* spp. (**75**) and *Comptonia peregrina* (bog-myrtle family, **76**) are inhabited by the actinomycete *Frankia* (**77**), while in the cycad *Macrozamia riedlei*, the nitrogen-fixing blue-green alga *Nostoc* is present in its surface roots. These varied micro-organisms utilize part of the fixed nitrogen for their own growth, but some is also available to the growing tree roots. Correspondingly, some of the sugars synthesized by the tree are used by the micro-organisms.

Symbiotic endomycorrhizae are commonly located within the fine roots of a wide variety of trees, including various species of broadleaved trees such as *Acer* (maple, sycamore), *Fraxinus* (ash), *Juglans* (walnut), *Magnolia*, and *Olea* (olive), as well as species of conifers such as *Juniperus* (juniper), *Sequoia sempervirens* (coastal redwood), and *Sequoiadendron giganteum* (giant sequoia). The root tips may develop some root hairs and usually there is little external evidence of the associated fungus; but within the root the fungi penetrate the cell walls of the outer parenchymatous tissue (cortex) to produce highly branched arbuscules (**78**). The thread-like mycorrhizae ramify through the soil and provide the tree with additional access to essential inorganic ions (especially phosphate) and water located in the soil some distance away from the nearest young roots.

Ectomycorrhizae are found in about 90% of northern hemisphere temperate broadleaved tree

74 Profuse development of root hairs on a leaf cutting of *Saintpaulia*.

75 Large warty nodules of the broadleaved tree *Alnus glutinosa* exposed by flooding from Loch Lomond, Scotland.

roots, and also in most conifers (Thomas, 2000). These mycorrhizae also occur in *Eucalyptus* and *Nothofagus* (southern beech) of the southern hemisphere and some tropical dipterocarps. Infected roots typically lack root hairs, are distorted, and may be coralloid-like (**79**). The fungus is visible

76 Large complexly lobed root nodule of the broadleaved shrub *Comptonia peregrina* grown in aeroponic culture.

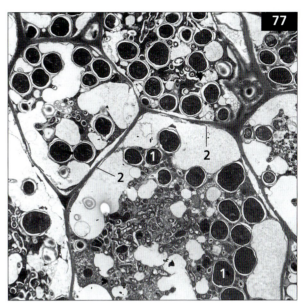

77 TEM showing vesicles (1) of the symbiotic actinomycete *Frankia* enclosed within the walls (2) of *Alnus glutinosa* root nodule cortical cells.

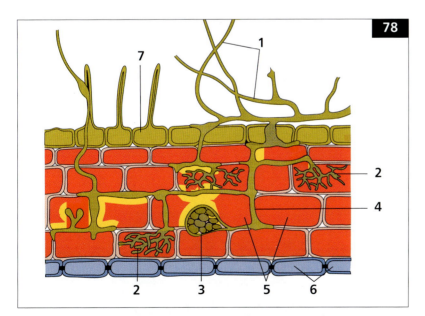

78 Diagram illustrating the internal structure of a tree root with endomycorrhizae. External hyphae in the soil (1), fungal arbuscle and vesicle (2, 3), fungal hypha (4) penetrating between root cortical cells (5), root endodermis (6), and root epidermis (7).

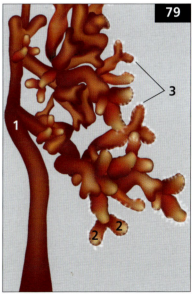

79 Diagram illustrating ectomy-corrhizae on a *Pinus* root (1), and the clumpy coralloid roots (2) which have developed in association with the investing symbiotic fungal hyphae (3).

externally on the root as a smooth sheath (**79, 80**), while internally its hyphae (fungal threads) ramify within the intercellular spaces of the outer cortical cells, but do not penetrate the cell walls. In late summer/autumn the fruiting bodies of many common mushrooms and toadstools represent the aerial extension of various ectomycorrhizal fungi (**81**). These are usually associated with a broad range of host conifers/broadleaved trees (Dix and Webster, 1995).

THE EVOLUTION OF LAND PLANTS AND TREES

The Earth is estimated to be some 4,600 million years old, with its crust beginning to develop about 4,200 mya (Willis and McElwain, 2002). Later, around 3,900 mya, as atmospheric water vapour began to condense, the earliest oceans appeared. Photosynthetic organisms first evolved in these seas about 3,500 mya, in the form of stromatolites, which are symbiotic mat-like communities of prokaryotic algae and bacteria. Their fossilized remains are abundant, and even today, living stromatolites occur in the shallow sea off the western coast of Australia (**82**), and in several locations elsewhere. Prokaryotes lack discrete nuclei and other membrane-bound organelles, but these organelles occur in the later-evolved eukaryotic organisms (**83**). The oldest known eukaryotic plant fossil is a large single-celled alga dating from about 2,100 mya. Filamentous red and green algae first appeared in the oceans about 1,200 and 800 mya, respectively (Willis and McElwain, 2002).

THE FIRST LAND PLANTS
SILURIAN PERIOD (443–417 MYA)

Soils had begun to form on the continental surfaces during the early Silurian period, due to the combined action of atmospheric weathering and the organic acids produced by lichens (symbiotic associations of fungi and algae) and micro-organisms growing on the exposed rock. During this period the first vascular plants began to occupy the margins of the Continental Shelf.

Cooksonia was an early colonizer, and its fossil remains occur in various European and North American sites. This leafless spore-bearing plant (**84**) grew only about 6 cm tall and bore sporangia on the

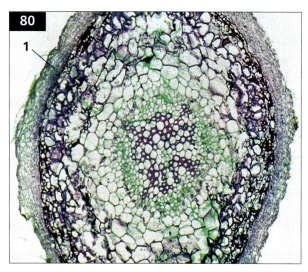

80 TS of a fine root of *Fagus sylvatica* (beech) invested by ectomycorrhizal hyphae (1).

81 Autumn fruiting bodies of the fungus *Amanita muscaria* (fly agaric), the underground mycelium of which forms an ectotrophic mycorrhizal association with tree roots, especially those of *Betula* spp. (birch). (Photo copyright of Norman Tait.)

82 Stromatolites exposed in the shallow coastal waters off the coast of Western Australia. (Photo copyright of Stephen D Hopper.)

tips of an equally branched (dichotomous) stem. Each sporangium contained many uniformly-sized spores (homospory), with each spore bearing three radiating ridges (trilete markings, **85**). Most specimens of *Cooksonia* are poorly preserved, but an epidermis containing stomata (pores for gaseous exchange) is evident in some. The underlying cortex, presumably photosynthetic, enclosed a core of elongated water-conducting cells (tracheids, Chapter 6).

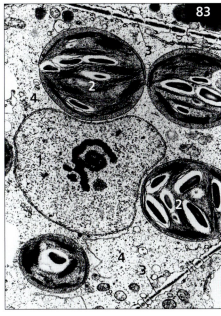

83 TEM of a green leaf cell of the moss *Polytrichum commune* (hair moss) showing the membrane-bound organelles characteristic of eukaryotic plants. Nucleus (1), chloroplasts (2), cell wall lined by plasmalemma (3), and ground cytoplasm (4).

84 Reconstruction of the late Silurian vascular plant *Cooksonia* (*ca*. 6 cm tall), a dichotomously branched and leafless early land plant with terminal sporangia.

85 Diagram of a spore tetrad (group of four haploid spores derived from the meiosis of a diploid spore mother cell) in an early vascular plant such as *Rhynia* and showing the trilete scars on a detached spore (top right).

34

Devonian period (417–354 mya)

A variety of somewhat larger homosporous land plants had evolved by the early-to-mid Devonian period. The Rhynie Chert from Scotland (86) contains the abundant silicified fossils of several such plants. *Aglaophyton* and *Rhynia* both formed upright, dichotomously-branched, and leafless stems up to 2 cm wide, with *Rhynia* reaching about 20 cm in height. Both plants arose from a basal rhizome bearing absorptive rhizoids (87), and their photosynthetic stems bore terminal sporangia (88). In *Rhynia*, the cylindrical core of the stem contained spirally thickened tracheids, but these were apparently absent in *Aglaophyton* (89).

86 Polished face of a block of Rhynie Chert, from which various primitive land plant remains may be identified in thinly sectioned material examined under the microscope.

87 Reconstruction of the leafless Devonian vascular plant *Rhynia* (ca. 50 cm tall). Note the terminal sporangia on the dichotomously branched shoots which arise from a rhizome.

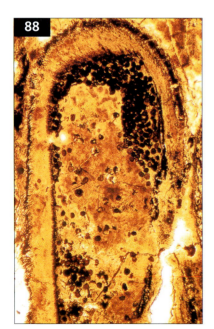

88 LS of the sporangium of a Rhynie fossil leafless plant, probably *Aglaophyton*.

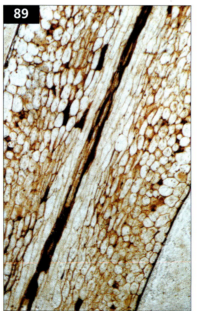

89 LS of the stem of a Rhynie fossil plant, probably *Aglaophyton*. Note the central strand of conducting cells.

90 Reconstruction of the leafy Devonian vascular plant *Asteroxylon*.

By contrast to these leafless forms, fossil lycopods, such as *Asteroxylon* (from the Rhynie Chert, **90**) and *Baragwanathia*, were densely covered with spiny leaves. Their stems contained a fluted central vascular (conducting) system, and in *Baragwanathia* this linked to veins in the leaves. The stem was up to 1 m tall and 1–2 cm wide in *Baragwanathia*, but was somewhat shorter in *Asteroxylon*. The present-day herbaceous lycopod *Huperzia selago* (**91**) is a distant relative of such ancient plants.

In the warm and humid climate of the mid-to-late Devonian period, extensive forests developed. These contained cone-bearing lycopods, such as the very common tree *Lepidodendron* (**92**), together with *Archaeopteris* (**93, 94**) and other progymnosperms.

91 Specimen of the extant herbaceous lycopod *Huperzia selago* (club moss).

92 Reconstruction of the Devonian/Carboniferous leafy arborescent lycopod *Lepidodendron* with its root system known as *Stigmaria*.

93 Reconstruction of the Devonian progymnospermous tree *Archaeopteris*.

94 Fossil specimen of *Archaeopteris* showing numerous leaves on thin side branches coming off from a main branch.

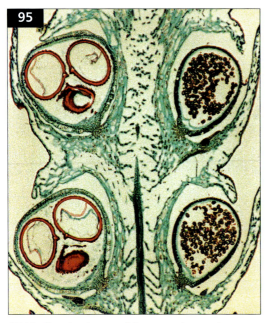

95 LS of a mixed cone of the extant lycopod *Selaginella*. Note the numerous small spores in the microsporangia and the few larger spores within the megasporangia.

96 Leafy twigs from a fine branch of *Lepidodendron*.

As in the present-day herbaceous lycopod *Selaginella* (**95**), many of these forms were heterosporous (with each sporangium bearing a few large spores or many smaller ones), while some reproduced by seed.

Lepidodendron was heterosporous, and some species apparently grew up to 35 m tall, with trunks 1 m in diameter (**92**). The upper trunk was repeatedly divided into finer branches to form a conspicuous crown, bearing simple, spiny leaves (**96**). After leaf abscission, the swollen corky leaf bases formed a characteristic interlinked diamond pattern on the trunk (**97**). These persistent leaf bases gave mechanical support to the tall *Lepidodendron* trunk. Internally, only scanty secondary wood developed, but the thick-walled outer cortex provided extensive additional support. The root system (often known as *Stigmaria*) was shallow but extended sideways up to 12 m from the base of the trunk (**98**), with the four primary branches dividing dichotomously and bristling with numerous side roots (**99**).

The mid-Devonian group of plants known as the progymnosperms are considered likely progenitors of modern seed plants (Willis and McElwain, 2002). *Archaeopteris* is a common fossil example and was

97 Coaly remains of a large branch of *Lepidodendron* showing the spiral, diamond-shaped pattern of its persistent leaf bases.

probably heterosporous. Its trunk reached up to 8 m in height by 1.5 m wide (**93**), and its roots penetrated down to 1 m beneath the soil surface. The upper trunk bore spirally arranged branches with abundant small leaves (**94**) and its wood resembled that of present-day conifers.

99 Root of *Stigmaria* (*Lepidodendron*) showing numerous lateral roots emanating from it in a bottle-brush arrangement.

98 A specimen of *Stigmaria* (the stump of *Lepidodendron*); this is one of a number of *in situ* stumps revealed by excavations on the Carboniferous site at Fossil Grove in Glasgow, Scotland.

100 Reconstruction of the Carboniferous arborescent horsetail *Calamites.*

101 Fossil frond of *Alethopteris*, a Carboniferous seed fern.

CARBONIFEROUS PERIOD (354–290 MYA)

Further tree groups had evolved by the early Carboniferous period, but the warm and wet equatorial forests were still dominated by *Lepidodendron* (92) and various other tree lycopods. The other major components of these forests were giant horsetail trees, such as *Calamites* (100), and some tree ferns (101), which bore seeds rather than spores. In more temperate regions, smaller arborescent lycopods and horsetails predominated, while ferns and seed ferns were common.

102 Specimens of the extant herbaceous pteridophyte *Equisetum* (horsetail). (Photo copyright of Norman Tait.)

103 Fossil stem of the Carboniferous arborescent horsetail *Calamites*.

104 Fossil of the leafy twig of *Annularia*, possibly belonging to *Calamites*.

Calamites was a common tree, with the main trunk growing up to 18 m high (**100**). The present-day herbaceous horsetails (**102**) are its distant relatives. The sporangia of *Calamites* were borne in cones, and some species were heterosporous. *Calamites* and other similar horsetails bore conspicuous rings of side branches with prominent longitudinal ribs (**103**), while their leaves were small and fused (**104**). Within the trunk and branches, a wide cylinder of wood surrounded a ring of vascular strands (veins) and central pith.

Various homosporous tree ferns inhabited these forests and were often very similar in appearance to present-day tree ferns such as *Dicksonia* and *Cyathea* (**6**). *Psaronius* was the largest Carboniferous tree fern, and grew up to 10 m high with an unbranched trunk bearing a terminal tuft of fronds several metres long. No woody thickening occurred in the trunk, but mechanical support was provided by its persistent woody leaf bases and the profuse development of prop roots. Seed ferns such as *Medullosa*, which grew up to 4 m tall with a trunk up to 50 cm wide, had a similar fern-like habit and foliage. Nevertheless, they reproduced by seeds rather than spores and their seeds are common Carboniferous fossils (**105**).

A number of conifer-like shrubs and trees bearing seeds evolved during the Carboniferous period.

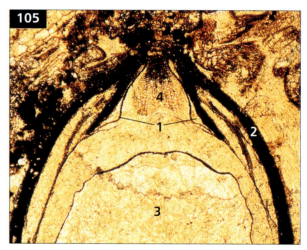

105 LS of an ovule from the Carboniferous seed fern *Lagenostoma ovoides*. Micropylar chamber (1), integument (2), female gametophyte (3), and nucellar beak (4).

Cordaites (**106**) had extensive woody thickening and grew up to 30 m tall, with a trunk of 1 m in width. The trunk branched frequently and bore spirally arranged, strap-shaped leaves up to 1 m long by 15 cm wide. The male and female cones were about 1 cm long but, unlike in modern conifers, the naked ovules (immature seeds) and pollen sacs were situated at the tips of the scale leaves. In other cordaitales, the leaves were much smaller and needle-like.

106 Reconstruction of the Carboniferous/Permian tree *Cordaites*.

107 Fossil specimen of the frond of *Nilssonia*, an arborescent Jurassic cycad.

108 Late Triassic fossil trunk of the conifer *Araucarioxylon* revealed by erosion of softer rocks of the Petrified Forest in the USA.

PERMIAN AND PERIODS TO THE PRESENT DAY

In the succeeding colder Permian period (290–248 mya), the tree lycopods and horsetails were no longer dominant (Willis and McElwain, 2002). However, there was an increase in various cordaitales, seed ferns, and primitive conifers, together with the newly evolved groups such as the glossopterids, bennettites, ginkgos, and cycads (**107**). The latter two groups persist into the flora of today (**15, 16, 46, 47**). During the subsequent Triassic period (248–206 mya), extensive evolution of new conifer lines occurred (**108**). *Araucaria* (**5, 21**), *Pinus* (**13, 18**), and *Taxus* (**49**) are some of their present-day relatives.

The earliest traces of flowering plants are found as pollen grains in the Cretaceous period (144–65 mya) in deposits about 130 million years old. The nature and form of these first flowering plants (angiosperms) is disputed, but they were probably herbs or small shrubs rather than trees (Willis and McElwain, 2002). By the late Cretaceous period, various present-day dicotyledonous tree genera such as *Ulmus* (elm), *Nothofagus* (southern beech), *Alnus* (alder), and *Betula* (birch, **8**) had evolved. Monocotyledons are rare in early Cretaceous deposits, but by about 110–100 mya, palms had evolved, and screw pines appeared early in the Paleocene period (65–55 mya). By the later Paleocene period, flowering plants had

109 A fossilized dicotyledonous tree leaf from the late Cretaceous period, from Mull, Scotland.

already become the dominant vascular plant group in most habitats (**109**), as they have remained to the present day (**1**).

SECTION 2 WORLD DISTRIBUTION OF FORESTS

CHAPTER 2

Northern boreal and montane coniferous forests

Aljos Farjon

INTRODUCTION

The boreal climatic zone occurs in the northern hemisphere and lies roughly in the latitudinal zone between the Arctic tundra to the north and temperate broadleaved forests and grasslands in the south. Its natural vegetation is dominated by forests of needle-leaved conifers, but various species of broadleaved trees such as *Betula* (birch), *Populus* (poplar), *Salix* (willow), and *Alnus* (alder) often occur in wetter areas. Here, the nitrogen-fixing actinomycete *Frankia* occurs in the roots of alder (Chapter 1). Under natural conditions, these boreal forests cover the region except where interrupted by swamps, lakes, river flood plains, high mountains rising above the climatic tree line, or by the oceans. Despite human impacts through logging and conversion, much of this broad forest belt still remains intact across North America (Canada and Alaska), northern Scandinavia, and Siberia.

The northern boreal forests have no counterparts in the southern hemisphere, not only due to the lack of great expanses of land at the appropriate latitudes, but also because the southern conifers belong to different taxa, which have different ecological strategies for survival. The northern conifer species have adapted to an extreme climate and have the ability to colonize vast areas while maintaining their dominance under these adverse conditions. Despite the numerous ice ages pushing all arborescent species far to the south, some trees were able to migrate again in warmer periods, such as exist at present, to reforest areas further north. However, many tree species remained at more southern latitudes or along ocean coasts, while some became extinct when retreat was not possible, as was the case in much of Europe.

Only a limited number of species had the necessary properties to enable them to re-colonize the boreal zone. In larger mountain systems (such as the Alps, Caucasus, Tien Shan, and Rocky Mountains), coniferous forests developed which were very similar to the boreal forests, but due to their isolation a number of different species could evolve in each of these mountain ranges. The Pacific coast has long functioned as both a refuge from the ice ages and a corridor for plant migration, with the result that its coniferous species differ from those in the interior of North America and Eurasia.

In this chapter a few examples of some important and characteristic species of the northern coniferous forests are described. These have been chosen to illustrate conifer diversity both taxonomically and ecologically.

CONIFERS OF THE BOREAL FOREST BELT

The coniferous forests of the interior continental boreal zone, known by the Russian term 'taiga', are dominated by a few genera and species in the Pinaceae (pine family). The most widespread of these are *Larix* (larch), *Picea* (spruce), and *Pinus* (pine), while *Abies* (fir) is more localized and usually montane. Others are essentially coastal conifers, or occur south of this zone. In both Eurasia and North America, each of these genera is represented in the boreal forests with generally no more than two, or at most three, species. Larches, the only deciduous conifers in this region, comprise two species (*Larix gmelinii* and *L. sibirica*) in boreal Eurasia and one (*L. laricina*) in North America. In mountains to the south of the taiga zone, other species such as *L. decidua* (European larch) and *L. occidentalis* (American larch) occur. The same paucity of species is seen in *Picea* (spruce), with both continents having only two. *Picea abies* (Eurasian Norway spruce) and *P. obovata* (Siberian spruce) actually integrate in the northeast of Russia. However, in North America, *P. mariana* (black spruce) and *P. glauca* are genetically and ecologically more distinct but occupy virtually the same geographical range.

There are a few more species of *Pinus* in the boreal zone, especially in North America, but no more than six in total. The most widespread of these is *Pinus sylvestris* (Scots pine), of which the Scottish populations are a tiny western outlier of a range that reaches from western Europe to the Pacific. In both continents, species belonging to soft pines (subgenus *Strobus*) occur along with the more common and widespread hard pines (subgenus *Pinus*), of which Scots pine is a representative. In North America, *Pinus banksiana* (Jack pine) is the most widespread of the pines, but in regions of the Rocky Mountains it is replaced by *P. contorta* (lodgepole pine). Whereas all conifers of the boreal zone become habitually stunted at the climatic limits of trees, certain species of pine are genetically determined to be shrubby, such as *P. pumila* (a soft pine with edible seeds) in eastern Siberia and *P. banksiana* in Canada, although the latter may grow to a small tree in more favourable soil conditions.

All these species of conifers must cope with a very short growing season and a prolonged, extremely cold winter. The taiga forest is under snow for up to eight months of the year, and often grows over permafrost or in water-logged soils. During the brief but hot and dry summers, it is subject to frequent forest fires, which may burn vast areas. Fires are difficult to control in the vast expanses of the largely road-less taiga; they determine the patterns of regeneration and forest composition almost as much as the limited growing season, and are responsible for a dynamic ecosystem with rapid biomass turnover. These conditions are, however, also the reasons why humans have largely left the northern boreal forests undisturbed. Habitation is difficult and agriculture impossible. Tree growth is slow, especially on poor soils away from river flood plains, resulting in thin, even stunted, trees that are uneconomical for man to exploit.

In the following sections some of the more important conifer species will be described briefly.

LARIX GMELINII

L. gmelinii (Dahurian larch) occupies all of eastern Siberia, east of a line from Lake Baikal to the mouth of the Yenisei River. Varieties of this larch described by various authors as different species occur, especially along the Pacific coast, and one variety forms an isolated population in central Kamchatka. The differences are mainly in the seed cones, all of which are smaller and have thinner scales than those of European or Siberian larch. Under favourable climatic and soil conditions this larch can reach a height of 35 m and a diameter of 1.5 m, and form a straight bole. However, it is often much shorter and curved with its branches spreading horizontally. As in all larches (**110**), the deciduous leaves grow on

110 Cones and foliage of *Larix decidua* (European larch). (Photo copyright of Bryan Bowes.)

short spur shoots, which increase only a few mm in length each year, and in autumn the leaves become bright yellow before falling. Among the boreal conifers, only larches are deciduous, but the fossil record shows that several other deciduous conifers, now confined to southern latitudes, once grew north of the Arctic Circle. Throwing off leaves in winter may have evolved in conifers in response to prolonged darkness, as a means of saving energy when photosynthesis has come to a standstill.

LARIX LARICINA

L. laricina (tamarack) in North America is the vicariant (i.e. a closely related sister species separated by a geographical barrier) of *L. gmelinii*. It is distributed across Canada, and in interior Alaska also has a large but disjunct population that some botanists consider as taxonomically distinct. The seed cones of *L. laricina* are even smaller than those of its Siberian counterpart (*L. gmelinii*), but otherwise they are very similar species. Morphologically, *L. laricina* is much more constant across its vast range than *L. gmelinii*, but it can grow to the same dimensions. Especially in the eastern parts of its range, it often occurs on peaty soils in swamps and muskegs, where it remains more stunted and is usually accompanied by *Pinus banksiana* and/or *Picea mariana*.

PICEA OBOVATA

P. obovata (Siberian spruce) is closely related to *P. abies* (Norway spruce), with the main morphological differences being the smaller seed cones and rounded scales in *P. obovata*. Intermediate forms are found in northeastern Russia, where the two taxa meet, and these are interpreted as natural (introgressive) hybrids. Siberian spruce extends from the Ural Mountains to the Sea of Okhotsk, but is absent from northeastern Siberia. Spruce stands tend to be more monospecific than larch stands, unless edaphic (soil) factors, such as on peaty soils, retard their growth. After fires, pioneer trees of *Salix* (willow), *Betula* (birch), and *Populus* (poplar) invade burnt areas, and are later succeeded by *Picea oborata*.

PICEA ABIES

P. abies (Norway spruce, **111**) occurs from Scandinavia and northern Russia to central Europe

111 *Picea abies* (Norway spruce) growing in the Alps of Switzerland.

and the Alps. In the history of the ice ages, when trees repeatedly retreated south and west over a time span of perhaps two million years, separation into varieties (perhaps even species) occurred. Hence, *P. abies* is much more variable morphologically than *P. obovata*, due to its more complicated history of retreat and separation. Growth forms of *P. abies* that were adapted to different climatic conditions evolved. In the interior of Eurasia, far from the ameliorating influence of oceans, the long winters are extremely cold and produce moderate quantities of dry snow. These conditions select for trees with slender, columnar crowns, and short, horizontally spreading branches. In much of Europe, winters are milder and produce large quantities of wet snow in the mountains. Here, spruce trees have longer, drooping, or even pendulous branches, allowing the snow to fall through. At times of maximum glaciation, *P. abies* was restricted to scattered refuges in southern and eastern Europe. When the climate warmed in interglacial times, the remnants spread out and often met. The resulting genetic diversity of *P. abies* is reflected in the extremely variable morphology of its seed cones. This variation occurs in both growth forms, as they were not subject to the same adaptive selection.

PICEA GLAUCA

P. glauca (Canadian or white spruce, **112**) extends all the way from Newfoundland to the coast of the Bering Sea in Alaska and, together with *P. mariana*, is one of the two spruces of the Canadian taiga. *P. glauca* occupies the better drained, often more fertile soils away from swamps and muskeg and, especially on river flood plains, it can attain a height

112 Forest of *Picea glauca* (white spruce) on the bank of the Porcupine River in the Yukon Flats, Alaska.

113 Natural stand of *Pinus sylvestris* (Scots pine) growing in the Scottish Highlands. (Photo copyright of Bryan Bowes.)

of 40–50 m with a trunk diameter of over 1 m. It develops the same narrow, columnar habit as *P. obovata*, and has small seed cones with thin, light-brown scales.

PICEA MARIANA

P. mariana (black spruce) and *P. glauca* virtually overlap in their ranges geographically, but are seldom found growing together. In boggy areas and where permafrost is prevalent, *P. mariana* replaces *P. glauca*, unlike in Siberia where there is only one species (*P. obovata*). Under such conditions, *P. mariana* forms an extremely slender tree, growing up to 30 m tall in the southeast of its range. Elsewhere, however, it is much smaller and is often dwarfed to a few metres. The small seed cones often aggregate at the top of the tree, where squirrels gnaw off shoots with cones and cause reiteration (i.e. secondary growth of twigs by activation of dormant buds). Hybrids between the two species, and with another spruce (*P. rubens*), have been reported, but the two species seem genetically well separated through most of their joint ranges.

PINUS SYLVESTRIS

P. sylvestris (Scots pine) has a very wide distribution (**113**) that reaches well beyond the boreal forest zone. Inside that zone, it is more restricted, but occurs in habitats ranging from dry sandy soils and rocky slopes to margins of moist bogs and lakes. It is found from Portugal to the Sea of Okhotsk in eastern Siberia. Beyond that line it is replaced by the shrubby species *P. pumila* (dwarf pine), although the two ranges partly overlap, with the dwarf pine forming the understorey

114 Upper trunk of *Pinus sylvestris* (Scots pine) showing its orange bark. (Photo copyright of Bryan Bowes.)

of the taller Scots pine. *P. sylvestris* is easily recognized by its orange, thin, papery bark, which occurs higher up the trunk (**114**) and on the branches of larger trees, its short, glaucous-green needles in fascicles of two, and its small cones with thin, dull-brown scales. In the Far East, some closely related species with similar characteristics occur. These may be only subspecies or varieties, a somewhat similar situation to that occurring with *Larix gmelinii*. The habit of *Pinus sylvestris* is dependent on its growth conditions. In dense stands of more or less even age, it becomes (like all northern conifers) slender with a straight bole, and such conditions are imitated in plantation forestry. In old, open stands it is more picturesque, as in some old forest remnants in eastern Norway and Scotland (**113**).

44

MONTANE CONIFERS

Conifer diversity increases substantially, especially in the uplands, south of the boreal zone and this trend continues down to subtropical latitudes. In this chapter, however, only conifers in montane regions, roughly down to 40° North, will be discussed. (This therefore excludes consideration of Mexico, the Himalayas, and most of China and Japan, all regions which are very rich in conifer species.) The montane regions have a more or less continuous contact with the boreal forests and served as refuges for many northern conifers during the ice ages. In these mountains there is a climatically determined zonation of vegetation that mimics the continental zonation from warm temperate to Arctic, but over greatly compressed distances.

This accounts for a substantial increase in diversity in these mountains. One of the most spectacular examples for conifers is the Sierra Nevada of California, rising from only a few hundred metres in the Central Valley of California to 4,418 m at Mt Whitney. The Eurasian mountains north of 40° are less species diverse, largely due to a less favourable E–W orientation, which blocked tree migration to the south in advance of the ice age winters. Species in the Pinaceae (pine family) generally dominate the forest but other conifers are important, with trees of the Cupressaceae (cypress family, including the formerly recognized family Taxodiaceae, redwoods) and, to a lesser extent the Taxaceae (yew family), enriching the forests. Among them are species with great tolerance to extreme situations and others that grow so well that they are the biggest tree species in the world.

ABIES ALBA

A. alba (white fir, **115**) is widespread in the mountains of central Europe, from the Pyrenees to the Carpathians. It is the only common species of fir in Europe, with other species only existing in relict populations in countries around the Mediterranean. On well-drained soils with abundant precipitation, *A. alba* grows as a tall tree (55–60 m) in a belt between the broadleaved and coniferous forests. Either it may form pure stands, or it is mixed with *Picea* and *Pinus* species, or often *Fagus sylvatica* (beech). In early summer, high in the crown of mature fir trees, seed cones are produced. These are

green and erect, and bear protruding (exserted) bracts. The cones turn brown and disintegrate in autumn. In this feature the genus *Abies* differs markedly from *Picea*, where the seed cones are pendulous and remain intact.

ABIES LASIOCARPA

A. lasiocarpa (subalpine fir) occurs throughout the Rocky Mountains from the Yukon Territory to New Mexico, from the subalpine zone to a maximum elevation of 3,500 m. Its slender, conical shape (**116**) is a characteristic sight in the western

115 *Abies alba* (white fir) in the Vosges Mountains of France.

116 *Abies lasiocarpa* (alpine fir) in the Rocky Mountains (Sawtooth Mountains) of Idaho, USA.

mountains of North America, where it grows in more or less open stands interspersed with alpine meadows. The densely set foliage branches and conical shape of *A. lasiocarpa* help to prevent breakage from the heavy snowfall the tree endures in much of its range. It produces erect, purple cones with hidden bracts.

PSEUDOTSUGA MENZIESII

P. menziesii (Douglas fir, which is not a true fir) occupies a range roughly similar to that of subalpine fir, but generally occurs at lower elevations. Two varieties of *P. menziesii* are commonly recognized, a coastal form and the inland form (var. *glauca*). The species extends into the mountains of Mexico but, being moisture-dependent, it becomes increasingly restricted southwards to favourable sites. On the coast in the Pacific Northwest, this conifer attains great size, to around 100 m tall and 4–5 m diameter. It is naturally mixed with other conifers in old-growth forest (**117**) but, unlike many other species of these forests, it has a capacity to colonize after disturbance in great numbers. After logging operations on old-growth stands, foresters have preferred to replant with Douglas fir. This has resulted in monospecific stands of this species – resembling managed plantations – in many mountain forests of western Canada and the USA. The seed cones of *Pseudotsuga* (**118**) differ greatly from those of *Abies*, and are of the pendulous, intact type (as in *Picea*) but with long three-pronged (trident) bracts above the rounded seed scales.

PINUS PONDEROSA

P. ponderosa (Ponderosa pine) is one of the most widely distributed pines in western North America. It occurs both in the Cascades and Sierra Nevada ranges, and throughout the Rocky Mountains, but not further north than 52° in British Columbia. Several forms or varieties are recognized and some botanists (but not the present author) include the mostly Mexican *P. arizonica* in it. There is also confusion about the 'typical' variety (var. *ponderosa*), which David Douglas apparently originally collected on the Columbia River in the north of its range. However, most foresters still believe that the magnificent trees found further south in Oregon and California constitute this typical variety. *P. ponderosa* is a 'sun-loving' pine, which forms open stands or may occur mixed with other

117 Large trunks of *Pseudotsuga menziesii* (Douglas fir) on the Pacific coast of Washington, USA.

118 Female cones of *Pseudotsuga menziesii*. (Photo copyright of Bryan Bowes.)

46

conifer species, as in the Sierra Nevada of California. Its bark forms characteristic large plates (**119**) and its egg-shaped seed cones are up to 15 cm long. These cones are generally smaller than those of its near relative *P. jeffreyi*, with which it can grow together in California. The clustered pollen cones of Ponderosa pine are strikingly coloured (**120**). There is a gradual shift in *P. ponderosa* from two-needled pines in the north, via two-to-three needled, intermediately located pines, to predominantly three-needled pines in the south. In general, needle numbers increase from north to south among many species of pine in North America (all those belonging to the subgenus *Pinus*), with two-needled pines in Canada and numbers of up to eight in Mexico.

JUNIPERUS OCCIDENTALIS

J. occidentalis (western juniper) occurs from Washington to California in the high mountains of the Cascade and Sierra Nevada ranges, although most of the tree-forming juniper species have a more southerly distribution. *J. occidentalis* is a magnificent tree (**121**), which grows extremely slowly but reaches 20 m in height and 2.5 m diameter. Its orange-red, stringy bark contrasts with dark-green, dense foliage of the minute scale leaves. Most trees are dioecious (bearing reproductive organs of one sex only), but exceptions occur. Western juniper is capable of survival under very extreme conditions, growing in cracks in bare granite rock with no soil content, and tapping seepage water that runs off the rocks into the fissure. As with all junipers, the wood is hard and extremely durable, but its slow growth makes commercial growing impossible. For this author, they are the most beautiful of all the conifers in the Sierra Nevada of California, which also has arguably the most beautiful conifer forests in the world.

COASTAL CONIFERS

At relatively northern latitudes, coniferous forests reach the coasts of both the Atlantic and Pacific

119 Trunk of *Pinus ponderosa* (Ponderosa pine) in California, USA.

120 Pollen cones of *Pinus ponderosa* (Ponderosa pine) shortly before pollen dispersal.

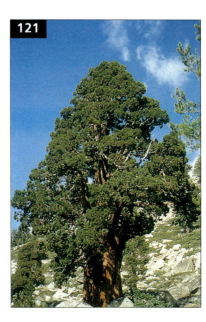

121 *Juniperus occidentalis* (western juniper) in the Sierra Nevada of California, USA.

Oceans. However, it is only on the Pacific coast that many conifer species are more or less restricted to a coastal strip, or to islands such as Japan. The coast of the North Pacific is a major refuge for ancient conifers that have become extinct elsewhere, their demise being due to geological history and to climate changes during the last few million years. Conditions for growth of conifers are probably nowhere better than in the Pacific Northwest of North America, but planted trees also thrive in Ireland and Scotland at the same latitudes, which indicates the similar climatic conditions in parts of Atlantic Europe.

The coniferous forests that extend from California to Alaska along the coast of the Pacific Ocean can be classified as temperate rain forest. The climate is continually moist or wet, with precipitation evenly distributed throughout the year as fog, rain, or wet snow, while temperatures are cool without extremes in summer or winter. These conditions prove to be particularly favourable for evergreen conifers, which attain extraordinary sizes. The forest begins directly above the line of highest ocean tides and climbs up to about 1,500 m in the south and only a few hundred metres in the north. There is usually a mixture of conifer species, with some conifers growing in groves as well as individually dispersed. Although some broadleaved (flowering plant) trees are common, especially along streams, they do not reach the canopies of the tall conifers. Ferns, bryophytes, and lichens are numerous and prominent, both on the forest floor and right up into the tops of the trees. Forest fires and storm damage are relatively rare and localized.

On the Asian side of the Pacific, conifers tend to grow more as solitary emergents above broadleaved trees, although conifer-dominated forests and groves occur on a smaller scale. Much of the old-growth forests have been replaced with other types of vegetation, as an agriculture-based civilization has had a much longer history here than in North America.

TSUGA HETEROPHYLLA

T. heterophylla (western hemlock, **122, 123**) is one of two species of *Tsuga* in North America (there are seven species in Asia, but none in Europe). The other species is *T. canadensis* (eastern hemlock), which is less coastal in its distribution. Hemlocks are extremely shade-tolerant, and are also the conifers that grow closest to the sea-coast in many areas. Their small leaves and numerous slender branches form dense horizontal sprays, blocking the sunlight from reaching the forest floor, which is usually covered in 20–50 cm of mosses, and hence is inaccessible. The tiny pendulous cones of the hemlocks disperse numerous small, winged seeds that can germinate on fallen trees where some litter has accumulated. Growing on these 'nurse logs' the hemlock seedlings start to compete for light, water, and nutrients. Eventually, a colonnade of trees will indicate where a long since decayed conifer tree once fell to the ground (**123**).

122 A 'nurse log' with numerous seedlings of *Tsuga heterophylla* (western hemlock) on the Pacific coast of Washington, USA.

123 Column of *Tsuga heterophylla* (western hemlock) trees standing on an old 'nurse log' in the rain forest of Washington, USA.

THUJA PLICATA

T. plicata (family Cupressaceae, **124**, **125**) is termed by American foresters, western red cedar. However, it is not a true cedar (*Cedrus*) as it does not belong to the Pinaceae. *T. plicata* has an eastern sister species in North America and a few relatives in eastern Asia, but the genus has become extinct in Europe. *T. plicata* is the largest species and has foliage with glossy green scale leaves, which are arranged decussately (i.e. in opposite pairs that are alternately shifted about 90° on the shoot) and diversified into laterals and facials. It bears erect, small, thin, and oblong-to-ovoid woody cones with spreading scales (**126**). The thin, stringy bark covers wood that can be split into large planks by a simple technique without the use of saws.

This allowed the native coastal Americans to build their large houses, as well as sea-going whaling boats. The wood is also rot-resistant, which is an important quality in a continuously wet climate. However, this property now works against long-term conservation of the species because it is still logged from old-growth forests to supply manufacturers of garden furniture, sheds, and greenhouses.

SEQUOIA SEMPERVIRENS AND SEQUOIADENDRON GIGANTEUM

Sequoia sempervirens (coast redwood, **127**) is famous for being the tallest species of all trees, with one specimen currently having reached 112 m in height. The species is now restricted to a narrow

124 *Thuja plicata* (western red cedar) on the Pacific coast of Washington, USA.

125 Three massive trunks of *Thuja plicata* (western red cedar).

126 Foliage and seed cones of *Thuja plicata* (western red cedar).

127 *Sequoia sempervirens* (coast redwood) on the Pacific coast of California, USA.

128 *Sequoiadendron giganteum* (giant sequoia), the champion 'General Sherman Tree' in Sequoia National Park, Sierra Nevada, USA.

129 Seed cone and foliage of *Sequoiadendron giganteum* (giant sequoia). (Photo copyright of Bryan Bowes.)

130 Bark of *Sequoia sempervirens* (coast redwood). (Photo copyright of Bryan Bowes.)

strip along the foggy Pacific coast, but before the ice ages it was once a common component of the northern coniferous forests of America and Eurasia. European settlers nearly logged it to oblivion, but it is one of the great achievements of the North American conservation movement that this did not happen, and that most of the remaining stands of redwoods are now protected. The closely related *Sequoiadendron giganteum* (giant sequoia, **128**) occurs in the Sierra Nevada of California, and was at one time considered as another species of the coast redwood. *S. giganteum* can grow enormous. One specimen, known as 'General Sherman', is thought to be the largest tree in the world based on its volume, and is 83 m tall with an estimated weight of 6,000 tonnes. Its congeners, or close relatives, are known from fossils as far away as Australia, which demonstrates an ancient lineage dating back to the early Cretaceous period more than 100 million years ago.

S. sempervirens has both scale and needle leaves, and bears small seed cones. *S. giganteum* has only scale leaves, but has larger cones (**129**) with more seeds in two rows on each scale. Both species have very thick, stringy, and fire-resistant bark (**130**), but persistent hot fires, caused by the ignition of accumulated dead wood, can creep inside hollow trunks and kill even the largest trees. *S. sempervirens* readily sprouts from the base (**131**) or from a large underground lignotuber, but this capacity for vegetative regeneration is relatively rare in conifers and absent in *S. giganteum*.

131 Clonal group of *Sequoia sempervirens* (coast redwood) growing in a Californian redwood forest. (Photo copyright of Bryan Bowes.)

CRYPTOMERIA JAPONICA

C. japonica (family Cupressaceae, **132**, **133**) is an endemic giant Japanese conifer related to the two sequoias of North America, as well as to *Taxodium distichum* (swamp cypress) to which it is more closely related. *C. japonica* is the only species in this genus, which is a common situation in the older and more 'basal' lineages of the family, due to extinction. The natural distribution of *C. japonica* in Japan is a contentious issue. A long and largely unrecorded history of exploitation of its durable wood has certainly destroyed virtually all old-growth forests. At the same time, however, it was widely planted in sacred temple grounds and protected groves or avenues, from where it may have spread again. More recently, this most important timber tree of Japan has been planted everywhere in forestry plantations, thus partly restoring its old range. It also grows in China but is probably not indigenous there, although it occurred across much of Eurasia before the ice ages. *C. japonica* has much in common with the sequoias in general appearance. Its cones (**134**), however, betray more affinity with the deciduous genus *Taxodium*, which is now confined to the eastern USA and Mexico, but was once abundant across the northern hemisphere.

CONCLUDING REMARKS

Conifers are dominant in northern boreal and montane forests, with several species of the Pinaceae (pine family) forming extensive stands covering vast tracts of sparsely inhabited land. These trees can successfully cope with the harsh conditions of a long and cold winter, and a short but intense growing season. They are the survivors of successive ice ages. Other species of conifer, often belonging to the Cupressaceae (cypress family), were as widely distributed before these periods of extreme cold, but have either become extinct or have retreated to small areas. In Europe, extinction was the rule rather than the exception and relatively few northern conifers remain to the present. Most of the conifer survivors are found around the coasts of the North Pacific, both in Asia and in North America. The forests of the Pacific Northwest, California, and, to a lesser extent, Japan, give an impression of what the northern coniferous forests once looked like across the globe. Their protection from unsustainable exploitation, and alteration by forestry practices, is a major task for the conservation of biodiversity on the planet.

132 Tall specimen of *Cryptomeria japonica* (Japanese red cedar) growing in an English botanic garden. (Photo copyright of Bryan Bowes.)

133 Trunk of *Cryptomeria japonica* (Japanese red cedar) planted tree in Ireland.

134 Cones of *Cryptomeria japonica* (Japanese red cedar).

CHAPTER 3

Temperate deciduous and temperate rain forests

Hugh Angus

TEMPERATE DECIDUOUS FORESTS (TDFs)

Temperate deciduous forests are characterized by their broadleaved trees, which shed their leaves in autumn (the fall) and form new foliage, or flush, in the following spring. The TDFs are principally located in four regions of the world.

- Eastern North America; here the forest extends from 30°N latitude in the south to 45°N latitude in the north, and inland from the east coast to longitude around 95°W.
- Western and Central Europe; apart from the breaks at the mountains of the Pyrenees, Carpathians, and the Alps, these forests form a nearly uninterrupted swathe extending from the Atlantic coast eastwards into Eastern Europe.
- Eastern Asia; this forms the third largest region of TDF and includes much of China and large parts of Japan. Its northernmost boundary is around 50°N latitude, and it extends southwards to around 30°N latitude and westwards to near the Tibetan Plateau.

- A smaller area of TDF occurs in the Near East around the Black Sea, Caucasus Mountains, Iranian Highlands, and the mountainous regions near the Caspian Sea. At its most westerly point it touches the southeastern edge of the European forest.

CHARACTERISTICS OF TDFs

The main characteristics of the temperate deciduous forest are its seasonality and the diverse range of flora and fauna found growing in association with the forest. No other ecosystem displays such seasonal changes, and is best defined as having hot summers and cold winters. The TDF year has two distinct phases: firstly, the vegetation period when most plant growth and reproduction takes place; and secondly, the dormant period when these processes stop or greatly slow down. Precipitation is generally in the region of 50–175 cm per annum. The rain falls throughout the year, but a greater percentage falls in the winter months. Temperatures can vary considerably. In most TDFs, however, it averages between a maximum of 18°C and a minimum of –3°C in the coldest months, while the average variation does not exceed 10°C in the hottest month.

Seasonality is mainly affected by climatic and day length differences, and there are four distinct seasons. In spring the temperature begins to rise, and the first tree leaves and flowers appear (**135**), although in some species flowers precede the leaves (**136**). By mid-summer most trees are in full leaf and have flowered (**137**, **138**). The longest day lengths occur in summer, coinciding with the greatest period of photosynthetic activity and leading to the main period of tree growth. By autumn the fruits have ripened (**139**), and the reductions in light and temperature levels trigger major changes in deciduous trees. Photosynthesis eventually ceases and the foliage begins to change colour (**140–142**) before leaf fall carpets the ground with dead foliage. This may sometimes accumulate (**143**), slowly rotting over succeeding seasons. During the drought of winter, the trees are in a dormant state and the leafless trees present a skeletal appearance (**144**).

135 Twig of *Fagus sylvatica* at bud break in springtime, England. This is native to the European temperate deciduous forest (TDF).

136 Small specimen of *Prunus spinosa* (sloe, blackthorn) native to the TDF of Europe – note how the tree in the English spring is smothered in white blossom before the emergence of its leaves. (Photo copyright of Bryan Bowes.)

137 Foliage of *Liriodendron tulipifera* (tulip tree) with the edges of its leaves characteristically indented. The species is native to the TDF of the eastern USA. (Photo copyright of Bryan Bowes.)

138 Large flower borne in summer on *Liriodendron tulipifera* (tulip tree).

139 In autumn a large crop of berries is ripe on this small tree of *Sambucus nigra* (elderberry) growing in England. Numerous species occur in the European TDF. (Photo copyright of Bryan Bowes.)

140 Senescent leaves of *Aesculus hippocastanum* in autumn just before leaf fall in Scotland. This species is native to the southern European TDF but is now naturalized in much of Europe. (Photo copyright of Bryan Bowes.)

141 Leaves of *Acer saccharum*, which contribute to the rich colours of the fall in the TDF of the eastern USA.

142 Fall foliage of *Sassafras albidum* from the TDF of the eastern USA.

143 A wood of *Fagus sylvatica* (beech) in Scotland at springtime with the beech leaves just bursting from their buds (flushing). Note the absence of shrubs and ground flora, and the dense layer of tree leaf litter. (Photo copyright of Bryan Bowes.)

144 Mature specimen of *Quercus robur* (English oak) leafless in winter and growing in open pasture in England. This species is also found elsewhere in the European TDF.

Several distinct vegetation layers are often apparent within a developing (**145**) or mature forest (**143, 146**). However, their component genera and species may vary considerably in the several world regions previously defined.

- Upper layer of deciduous tree species (**143, 146**), dominated by specimens usually in excess of 15 m tall at maturity, and sometimes growing to 30 m or more (*Table 1*). This layer is often composed of varying species of the Fagaceae (beech family, **143, 144**), *Castanea* (sweet chestnut), *Acer* (maple), *Carya* (hickory), *Fraxinus* (ash), *Juglans* (walnut), *Tilia* (lime, basswood), and *Ulmus* (elm).

- Small tree layer made up of both saplings from the larger trees (as above) and species of less tall trees. These trees will vary somewhat in different regions, but examples are species of maple, such as *Acer palmatum* (Japanese maple, **147**), *Amelanchier* (serviceberry), *Alnus* (alder), *Betula* (birch, **145**), *Cornus* (dogwood), *Sambucus* (elder, **139**), *Sassafras* (**142**), and *Oxydendrum* (sourwood).

- Shrub layer composed of various smaller woody species. Some common constituents are *Corylus* (hazel), *Hamamelis* (witch hazel), *Rubus* (bramble), *Rhododendron* (**148**) and other members of the heather family such as *Erica* (heather), and *Vaccinium* (cranberry, blueberry).

Table 1 The most important tree families, native genera, and species represented in the three largest areas of temperate deciduous forest

Genus	Family	Species in eastern North America	Species in Europe	Species in East Asia
Acer	Aceracaea	10	9	66
Alnus	Betulaceae	5	4	14
Betula	Betulaceae	6	4	36
Carpinus	Carpinaceae	2	2	25
Castanopsis	Fagaceae	0	0	45
Cyclobalanopsis	Fagaceae	0	0	30
Diospyros	Ebenaceae	1	0	25
Fagus	Fagaceae	1	2	7
Fraxinus	Oleaceae	4	3	20
Lithocarpus	Fagaceae	1	0	47
Magnolia	Magnoliaceae	8	0	50
Malus	Rosaceae	1	1	8
Populus	Salicaceae	4	4	33
Prunus	Rosaceae	3	4	59
Quercus	Fagaceae	37	18	66
Salix	Salicaceae	13	35	97
Sorbus	Rosaceae	3	5	18
Tilia	Tiliaceae	4	3	20
Ulmus	Ulmaceae	4	3	30

145 This stand of *Betula* (birch) is a pioneer establishment of young trees growing on peaty moorland in Scotland. (Photo copyright of Bryan Bowes.)

146 *Fagus grandifolia* woodland in the TDF of the eastern USA in the fall (autumn).

147 The attractive foliage of *Acer palmatum* (Japanese maple) growing in an English arboretum, although the tree is native to Japan and China. (Photo copyright of Bryan Bowes.)

148 Winter view of an English mixed wood showing the trunks of several young trees, leaf litter, and small brushwood on the ground, and a shrub layer of *Rhododendron*. (Photo copyright of Bryan Bowes.)

- Herbaceous (field) layer made up primarily of perennial plants, which often flower before the trees come fully into leaf (**149–151**). Some examples are *Anemone* (wood anemone), *Dicentra cucullaria* (Dutchman's breeches), *Ranunculus* (buttercup), *Hyacinthoides non-scripta* (bluebell, **149**), *Mercurialis* (dog's mercury, **150**), *Oxalis* (wood sorrel), *Glechoma hederacea* (ground ivy, **151**), *Hepatica*, *Sanguinaria* (bloodroot), *Trillium*, *Viola* (violet), and various ferns. Many of these species die down before or during winter (see also Chapter 10).
- Ground layer sometimes covered with very dense leaf litter (as in some beech woods, **143**) or small brushwood (**148**), but otherwise made up of various mosses (**152**), liverworts, lichens, and fungi.

The range of species within any layer of forest is considerable. In colder regions there tend to be less broadleaf deciduous species, but a greater percentage of conifers, which are typically evergreen and retain their leaves for several years. This is most obvious in the more northerly areas of the TDFs, or at higher elevations in the more mountainous regions. By contrast, in the warmer areas of the south, there is a greater percentage of evergreen broadleaf species.

Soil fertility is also an important factor, with higher-nutrient soils supporting a greater diversity of trees and other plants. Most temperate deciduous forests are found on nutrient-rich soils, and hence often have to compete with land cleared for growing crops and farming. The complex mix of the above

149 Extensive swathe of *Hyacinthoides non-scripta* (bluebell) in spring, forming a typical part of the TDF ground flora in a British wood.

150 Densely crowded specimens of *Mercurialis perennis* (dog's mercury) in England – a woodland TDF herb flowering in early- to mid-spring. (Photo copyright of Bryan Bowes.)

151 The woodland TDF perennial *Glechoma hederacea* (ground ivy) flowering at springtime in England. (Photo copyright of Bryan Bowes.)

152 A mixed wood in Scotland (in winter) showing a dense covering of moss on both the ground and branches of a fallen tree – note the biennial herb *Digitalis purpurea* (foxglove) among the moss carpet. (Photo copyright of Bryan Bowes.)

factors often leads to a high degree of diversification within an individual forest stand. Also, the original forest trees have often been replaced by non-native species, which show better growth rates and yield a higher crop of timber. In particular, broadleaved species have often been replaced by conifers.

Nevertheless, despite all such variations, various large trees that are members of the Fagaceae (*Table 1*) still predominate. Five such genera (*Castanopsis*, *Cyclobalanopsis*, *Fagus*, *Lithocarpus*, and *Quercus*) collectively comprise 254 species within the TDF. Of these, *Quercus* (oak, **144**) is represented by 121 species, with 18 in Europe and 37 in eastern North America. Despite there only being 10 species of *Fagus* (beech, **135**, **143**), it is a very important constituent of these forests, and often forms large stands of beechwood (**143**, **146**), particularly in Europe. *Fraxinus* (ash) has 27 species, with the greatest number of species in eastern Asia. However, it is also an important tree in Europe (**153**). *Salix* (willow) has the largest representation in the TDF, with 145 species (*Table 1*), although most are small trees or shrubs.

CHARACTERISTIC EASTERN NORTH AMERICAN TDF TREES

The deciduous forests of this region remain among the most diverse and magnificent of any in the world (**141**, **142**, **146**). The main deciding factor in their creation is the humid continental climate found in this area of North America. Their great species diversity is in part due to a more or less natural forest cover having been retained in some areas, despite some modification by logging. However, elsewhere the TDF has a relatively long history of being managed by man. Transition forest is associated with either large-scale logging, or the natural reversion of trees becoming re-established on abandoned agricultural lands (see also Chapter 10), to form a forest with a more or less natural composition. The characteristic tree species within forest stands vary greatly depending on the local topography, soil, and the amount of human disturbance. In the north, the TDF gives way to conifer forest, while to the west the trees are replaced by grassland. In the south, where growing seasons are longer, a greater percentage of evergreen species appear. Some examples of large overstorey trees, with a widespread distribution in the region, are described below.

Acer saccharum (sugar maple, **141**) is the tree from which maple syrup is derived. Its sap contains up to 3% sucrose, and in winter to early spring the tree can be tapped; the exudate is then concentrated by boiling to provide the commercial syrup. The tree is one of the more dominant deciduous trees and may grow up to 35 m tall. In some areas it can account for up to 80% of the forest cover. *A. saccharum* is a shade-tolerant species which thrives where the soils are light in nature and well drained.

Fagus grandifolia (American beech, **154**) grows up to 25 m or so tall and is a highly shade-tolerant species which can form nearly pure stands in some areas (**146**). It grows best in the eastern part of the region where the soils tend to be moister.

153 Midsummer foliage of *Fraxinus ornus* (manna, flowering ash) growing in an English arboretum but native to the TDF of southern Europe and western Asia. (Photo copyright of Bryan Bowes.)

154 Autumn foliage of *Fagus grandifolia* (American beech) just before leaf fall.

155 Foliage of *Quercus coccinea* (scarlet oak), a native TDF tree of the central and eastern USA. (Photo copyright of Bryan Bowes.)

156 Foliage and fruits of *Aesculus pavia* (red buckeye), a small native TDF tree of the eastern USA. (Photo copyright of Bryan Bowes.)

The shade-tolerant seedlings can survive for long periods in deep shade until light levels increase. When gaps in the forest canopy occur, some saplings can grow out to develop into large trees. *Betula alleghaniensis* (yellow birch) is a valuable timber tree and reaches up to 30 m high. It grows best in very wet areas, and its shade-tolerance, range, and population are more limited than sugar maple and beech.

Populus tremuloides (quaking aspen) is an important component of the northern part of the deciduous forest. This species is shade-intolerant, grows on wet soils over a wide area of North America, and often forms extensive clonal forest stands. The individual tree trunks are often interlinked by a common root system, which originates from numerous root suckers and natural root graftings. However, the aspen may also grow in association with other tree species. *Liriodendron tulipifera* (tulip tree **137**, **138**) is a fast growing, beautiful tree, which reaches up to 50 m in height. It is present over much of the region and extends southwards to Florida. Its large leaves have characteristic indented edges (**137**), while its large (up to 6 cm long) tulip-shaped flowers are also very distinctive (**138**).

Other larger deciduous tree species occur but in smaller numbers and, while sometimes being locally abundant, are generally not so widespread through the TDF. These include: *Quercus coccinea* (scarlet oak, **155**) and *Q. rubra* (red oak); *Tilia americana* (American basswood) and *T. heterophylla* (white basswood); *Fraxinus nigra*; *Acer rubrum*

157 Fall foliage of a species of *Carya* from the eastern USA.

(red maple) and *A. saccharinum*; *Aesculus flava* (yellow buckeye) and *A. pavia* (red buckeye, **156**); *Liquidambar styraciflua* (sweetgum); several species of *Carya* (hickory, **157**); and *Robinia pseudoacacia* (false acacia, **158**). *Castanea dentata* (American chestnut) was also an important tree in the eastern Appalachians until virtually eliminated by chestnut blight, which was accidentally introduced from China early in the 20th century. Smaller species include *Sassafras albidum* (sassafras, **142**), *Oxydendrum arboreum* (sourwood, black locust), and *Acer pennslyvanicum* (moosewood).

Various large conifers also occur, including *Tsuga*

158 Foliage of *Robinia pseudoacacia* (false acacia), a native TDF tree of the central and eastern USA. (Photo copyright of Bryan Bowes.)

159 An arboretum specimen of *Tsuga canadensis* (eastern hemlock) growing in England but native to eastern North America. (Photo copyright of Bryan Bowes.)

canadensis (eastern hemlock, **159**), which grows best on damp acid soils and is an important component within the deciduous forest. It is most plentiful towards the west. At higher elevations it is often found in protected situations, such as valleys or on north-facing slopes. *Pinus strobus* (white pine) is another significant forest tree, and grows best under drier conditions in the northern and western areas. It readily colonizes ground cleared by fire and can be considered a fire-dependent species.

CHARACTERISTIC EUROPEAN TDF TREE SPECIES

Although providing a fertile environment, the forests of Europe do not contain the species richness of the North American or Asian deciduous forests. Although the warming influence of the Gulf Stream has allowed the European deciduous forest to extend a good 10 degrees of latitude further north than is the case for either the eastern North American or the Asian TDFs, it does not extend so far south as in the latter regions. In European forests, postglacial history, climate, and human disturbance have been the greatest factors in limiting diversity, and the forests are frequently dominated by *Fagus sylvatica* (beech, **143**).

In Continental Europe most mountain ranges run in an east–west direction. This has had a three-fold effect on the natural vegetation of the deciduous forest: the obstruction to the migration of trees and other plants northwards; the reduction of the warming influence from the more southerly Mediterranean climate; and moist air masses blowing into the

Continent from the Atlantic, which can penetrate further eastwards into Central and Eastern Europe.

Two main large overstorey tree genera, *Fagus sylvatica* (common beech, **135, 143**) and *Quercus* (oak) are found throughout the European deciduous forest.

Fagus sylvatica has by far the greatest distribution within the more northerly TDFs, but from Bulgaria and to the southeast it is replaced by *F. orientalis*. Beech grows best on drier soils but also tolerates varying soil types well. It is shade-tolerant and in certain areas forms dense stands where very few plants can grow underneath the canopies (**143**), but its own seedlings are very shade-tolerant (Chapter 10).

Oak (*Quercus*) is a very important genus throughout the whole of this region. It generally grows best in damper soil conditions than beech, and is usually found among a much greater diversity of other trees. In conditions where beech trees do not thrive, oaks tend to be their main replacements. However, as oaks generally grow on the better soils, they tend to be more disturbed by human settlements. In the north of the region the predominant species are *Quercus robur* (English oak, **144**) and *Q. petraea* (sessile oak). Further south, other species such as *Q. cerris* (Turkey oak), *Q. frainetto* (Hungarian oak), and *Q. ilex* (evergreen oak) appear with increasing frequency.

Other important, large broadleaf tree species occur in smaller numbers. These include *Aesculus hippocastanum* (**140**); *Carpinus betulus* (hornbeam);

Fraxinus excelsior and *F. angustifolia* (ash); *Acer pseudoplatanus* (sycamore, **160**); *Tilia cordata* (small-leaved lime, **161**) and *T. platyphyllos*; and *Castanea sativa* (sweet chestnut).

Various smaller tree species are also frequently present; for example *Acer campestre* (hedge maple); *Alnus glutinosa* and *A. incana* (alder); *Betula pubescens* and *B. pendula* (birch, **145**); *Corylus* (hazel); and *Salix* (willow).

In certain areas, particularly in the transition zones from deciduous to the more northerly boreal forest region, several conifers form an important evergreen component of the TDF. These species are typically confined to the colder and higher elevation sites (Chapter 2), with the main species including *Abies alba* (silver fir), *Picea abies* (Norway spruce), and *Pinus sylvestris* (Scots pine).

CHARACTERISTIC EAST ASIAN TDF TREE SPECIES

This forest type is probably the most species diverse found in any TDF region. However, prolonged and extensive human influence has turned most of the formerly, very extensive deciduous forest into farmland. Today, only relatively small areas of true forest are still in existence. These occur mainly in southwest and northeast China, but some also remain in Korea and Japan. The different locations have quite different land forms, which further adds to the forest diversity. In the mountains, the lower slopes are sometimes covered with sub-tropical rainforest. At higher levels, deciduous broadleaved forest is present, which gives way to montane coniferous forest near the timberline.

A huge number of tree species occurs in such forests, and only the main genera will be considered here (*Table 1*). Four principal genera of the Fagaceae (beech family) are present, with 66 species of *Quercus*, 47 species of *Lithocarpus*, 45 species of *Castanopsis*, and 30 species of *Cyclobalanopsis*. Four other tree genera – *Fraxinus*, *Tilia*, *Acer*, and *Populus* – are represented by 139 species, with many forming part of the overstorey. *Cercidiphyllum japonicum* (Katsura tree, **162**) is the

160 Mature specimen of *Acer pseudoplatanus* (sycamore) in Scotland seen without foliage in winter. This species is native to the TDF of southern and central Europe but, from its introduction to Britain in the 16th century, has now become naturalized. (Photo copyright of Bryan Bowes.)

161 Part of an extensive stand of *Tilia cordata* (small-leafed lime) growing in an arboretum in England, but also found elsewhere in the European TDF. The individual coppiced specimens of this stand are all clonal, and are believed to be derived from a single specimen growing some 2,000 years ago. (Photo copyright of Bryan Bowes.)

162 Foliage of *Cercidiphyllum japonicum* (Katsura tree) growing in an English arboretum but native to Japan and China. (Photo copyright of Bryan Bowes.)

single representative of the family Cercidiphyllaceae, and grows up to 30 m tall in Japan and western China. Conifers are also among the larger trees, particularly at higher elevations. They include species of *Larix* (with *L. kaempferi* indigenous to Japan), *Pinus*, *Picea*, and *Tsuga*. The understorey is also rich, containing *Abies koreana* (indigenous to Korea, **163**) and species of *Carpinus*, *Acer* (**147**), *Corylus*, *Dipteronia*, *Tetracentron*, and many other genera.

TEMPERATE RAIN FORESTS (TRFs)

The temperate rain forests, in which numerous broadleaf and conifer tree species grow, are scattered into several relatively small areas of the northern and southern hemispheres:

- Western North America, from latitudes of around 60°N to 40°N. This relatively narrow belt of land hugs the Pacific coast and extends northwards from California to southern Alaska.
- South America (at its southern tip), extending from latitude of around 38°S in southern Chile to 55°S in Tierra del Fuego.
- Tasmania (Australia), forming a relatively small area concentrated along the western edge of the island.
- Other small areas of temperate rain forest include parts of Norway, the UK, Turkey, and New Zealand. Some of these remnants are very small and fragmented, often because of man's influence.

163 Specimen of *Abies koreana* showing its prominent dark blue female cones – this fir is native to Korea. (Photo copyright of Bryan Bowes.)

CHARACTERISTICS OF TRFs

Temperate rain forests are among the most threatened and rare forest habitats in the world. It is believed that of the original 30–40 million hectares, only some 44% remains intact and free from logging activities. TRFs occur in a broad range of latitudes but cover relatively small areas and are all concentrated in thin coastal strips (Chapter 2). The main factor in the creation of these rain forests is their proximity to both mountains and oceans, which combine to produce a high rainfall and equable climate. The forest structure is diverse and, due to the wet conditions, forest fires are infrequent. The rainfall is typically over 1,400 mm per annum but can be less in some areas. However, even in these areas, the dense summer fogs give a constant supply of water to support forest growth. By contrast, rainfall in some areas may exceed 3,500 mm per year, a good example being the Olympic Peninsula of western North America.

The year-round growth in temperate rain forests allows trees to reach a large size quickly, and such forests are among the most productive in the world today. This often creates environmental and sustainability problems, since the conifers in particular are highly prized for logging by the forest industries. Although very diverse, these forests are often dominated by evergreen conifers, particularly in North and South America.

There are four major subgroups of temperate rain forest based on temperature and precipitation patterns (Alaback 1990; Weigand, 1990).

- Boreal TRF has the lowest number of growing days, a mean July average temperature of 12°C, and considerable snowfall. The Alaskan peninsula and the most southerly tip of Chile are good examples.
- Perhumid TRF has a mean July temperature of 12–16°C and minimum snowfall. Vancouver Island in Canada and the Isla Chiloe in Chile are prime examples.
- Seasonal TRF has a mean July temperature of 16–25°C and minimal snowfall. Central Washington State (USA), Chile, and Tasmania provide the best examples.
- Perhumid subtropical TRF has a mean July temperature of greater than 25°C and virtually no snowfall. The redwood forests in the Californian coastal fog belt are an excellent example.

Characteristic western North American TRF tree species

Here, the temperate rain forest lies on the coast sandwiched between the Pacific Ocean and the Rocky Mountains. It is the largest area of such forest in the world, extending north–south for more than 2,400 km and covering between 14 and 20 million hectares. The proximity of the Rocky Mountains gives protection from severe weather as well as trapping the ocean moisture. The mean annual temperature varies between 4°C and 12°C. Temperatures rarely exceed 30°C in summer or drop below freezing during the winter.

These equable temperatures frequently support the growth of a great variety of flora and the production of very large and old trees. Since the soils are generally wet, and have high nutrient levels, the trees do not root deeply. This leads to frequent wind throw and results in the creation of habitats which are particularly good for the growth of epiphytic lichens, mosses, and ferns (**164**). Fungi are also prevalent in large numbers. Nurse logs are a characteristic of the Olympic Peninsula, where tree seeds (and other species) germinate in the humus on a rotting log (Chapter 2). The saplings wrap their roots around the log and grow down into the soil, and some eventually grow to mature trees (**165**). Each year the coastal temperate rain forests accumulate up to 500–2,000 tonnes of organic material per hectare (Juneau and Fujimori, 1971; Franklin and Waring, 1980; Alaback, 1989). This can be as much as five times the amount accumulated in tropical rain forests.

This western North American rain forest well demonstrates a range of the different subgroups mentioned above; perhaps the most interesting of these is the Coastal Redwood belt in the south. Although its yearly rainfall is low, dense coastal fogs sweep in from the Pacific, so that the actual moisture levels remain high and support this type of habitat. Although a variety of trees occurs, the populations are dominated by a relatively few massive conifer species (Chapter 2), which often exceed 40 m in height and form the world's most valuable natural timberlands. However, much of the original natural rain forest has been logged, and many of these native conifer species are now grown for timber in other parts of the world.

Picea sitchensis (Sitka spruce, **166**) is found throughout the whole of the rain forest belt except in

164 Moss/lichen-laden tree branches in the TRF in Oregon, USA.

165 Wind-blown prostrate tree trunk ('nurse log') with two large adventitious tree species growing astride it, in the coastal TRF of Oregon, USA.

166 Detail of a twig of *Picea sitchensis* (Sitka spruce) growing in Scotland. This species is, however, native to the TRF of western North America. (Photo copyright of Bryan Bowes.)

167 Specimen of *Pseudotsuga menziesii* (Douglas fir) growing in Scotland. This species is, however, native to the TRF of western North America. (Photo copyright of Bryan Bowes.)

168 Fruits of *Aesculus californica* (Californian buckeye) growing in England. This species is, however, native to the TRF of western North America. (Photo copyright of Bryan Bowes.)

the very south. It is a most valuable timber tree and grows to 50 m and more tall, with a diameter up to 1.5 m. Its high-grade timber has been used for building, boat building, and piano sounding boards, but it is also pulped to produce newsprint. Nowadays, Sitka spruce is the main tree plantation species grown in Britain.

Pseudotsuga menziesii (Douglas fir, **167**) is another important conifer of these forests. It is a very tall species (up to 60 m in height) and grows best in a free-draining soil. Its timber has a rich red colour with white banding, provides extremely valuable lumber, and is also often used as a veneer on plywood. Its specific name is derived from the Scottish explorer Archibald Menzies, who discovered the tree in 1793, although its common name is derived from another Scot who first sent the seeds to Europe in 1827. Various other native conifers, such as *Abies amabilis* (Pacific silver fir) and *Chamaecyparis nootkatensis* (Nootka cypress), are also important to the timber industry. Several others, such as *Tsuga heterophylla* (western hemlock), *Thuja plicata* (western red cedar), and *Sequoia sempervirens* (coastal redwood), have already been described in more detail in Chapter 2.

Various broadleaf species also grow in the region. *Populus trichocarpa* (black cottonwood) is a large tree and is found in small quantities throughout the area. Other trees include *Acer macrophyllum* (big leaf maple) and *Acer circinatum*, *Alnus rubra*

(red alder), *Chrysolepis chrysophylla* (golden chinkapin), *Lithocarpus densiflorus* (tanbark oak), and *Platanus racemosa* (western sycamore). Assorted species of *Quercus* (oak) occur, with numerous specimens of *Q. agrifolia* (Californian live oaks) and *Q. chrysolepis* (canyon live oak) often growing in open groves. These sometimes occur along with small trees of *Aesculus californica* (Californian buckeye, **168**).

The acorns of *Q. agrifolia* were particularly prized for food by the native North Americans, and were ground to a meal which, after leaching to remove its bitter taste, was baked for bread. Likewise, the large and poisonous seeds of buckeye were first boiled in water to extract the toxin and then ground into flour. However, the untreated flour was sometimes thrown into pools to stupefy the fish, which were then easily caught.

CHARACTERISTIC SOUTH AMERICAN TRF

TREE SPECIES
The rain forest is located in southern Chile, where it covers an area of 7–11 million hectares and extends up to over 600 m above sea level. The forest has a much greater mix of broadleaf and conifer tree species than the rain forest of North America, and woody vines are also frequently found. Many of its native tree species are very threatened due to over-exploitation of the forests for timber and clearing for agriculture. The harvested trees are

used mainly for local construction and fuel, while wood chips are produced mostly for export. Although in a number of areas the forests are still threatened, the creation of 25 national parks and 14 national reserves now provides some protection.

Nothofagus (southern beech) is the most widely distributed tree genus, and its various species are also among the more common of the rain forest. Most species are important timber trees and include *Nothofagus obliqua* (Mirbel, **169**), which once formed extensive natural forests, *N. dombeyi*, *N. betuloides*, *N. nitida* (**170**), *N. nervosa*, and *N. pumilio*. Other species of *Nothofagus* are native to Australasia (see below), giving an indication of the former distribution of this genus in Gondwanaland

169 Detail of the foliage of *Nothofagus obliqua* (southern beech) from the southern Chilean TRF. (Photo copyright of Martin Gardner.)

170 Foliage of *Nothofagus nitida* from the TRF in southern Chile. (Photo copyright of Martin Gardner.)

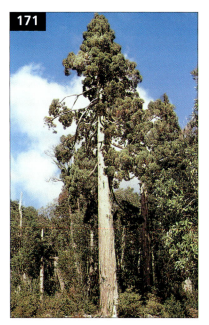

171 Mature specimen of *Fitzroya cupressoides* from the TRF in southern Chile. This species is endangered due to over-logging. (Photo copyright of Martin Gardner.)

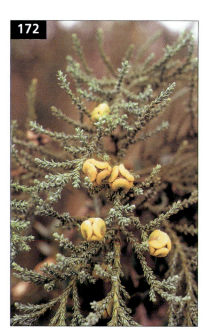

172 Twigs and cones of *Fitzroya cupressoides*. (Photo copyright of Martin Gardner.)

before the break-up of this supercontinent by continental drift.

Fitzroya cupressoides (Alerce, **171**, **172**) was formerly widespread in the rain forest, but now the trees have mostly been harvested for their valuable timber to the point of near extinction. *Araucaria araucana* (Chile pine, monkey puzzle, **173**, **174**) is a valuable timber tree restricted to the northern zones of the rain forest; again today, it is threatened due to over-logging in the wild. The tree also provides a valuable food source for the local people, but the seeds take two years to ripen on the female cones (**174**).

Saxegothaea conspicua (**175**) is an important monotypic species, while various trees, including *Aextoxicon punctatum* (**176**) and *Pilgerodendron*

173 Mature specimens of *Araucaria araucana* in the TRF of southern Chile. This species is also endangered due to over-logging, and is now a protected species in Chile. (Photo copyright of Martin Gardner.)

174 Female cones of *Araucaria araucana*. (Photo copyright of Martin Gardner.)

175 Foliage of the conifer *Saxegothaea conspicua* from the TRF in southern Chile. (Photo copyright of Martin Gardner.)

176 Detail of the foliage of *Aextoxicon punctatum* from the TRF of southern Chile. (Photo copyright of Martin Gardner.)

177 Flowers of the arborescent *Weinmannia trichosperma* from the TRF of southern Chile. (Photo copyright of Martin Gardner.)

178 Flowers of the arborescent *Eucryphia cordifolia* from the TRF of southern Chile. (Photo copyright of Martin Gardner.)

uviferum, are native to the area. Other abundant woody plants growing in this rain forest include *Laurelia sempervirens*, *Weinmannia trichosperma* (177), *Eucryphia cordifolia* (178), *Drimys winteri*, *Podocarpus nubigenus* (179), and *Hydrangea integerrima*.

CHARACTERISTIC TASMANIAN TRF TREE SPECIES

Although small pockets of TRF occur on the Australian mainland, by far the largest areas are found in Tasmania, where TRF now covers some 10% of the state. In the last century, over 7% of Tasmania's rain forest was cleared and either planted with trees that produce a high yield of timber, or used for farming. The remaining areas of temperate rain forest contain species representative of some of Australia's most ancient flora, and the trees show a high degree of endemism. The forest contains a good mix of both deciduous and coniferous trees but, unlike the South American TRF, woody climbing plants are rare. Some of the more important trees are described below.

Nothofagus cunninghamii (myrtle beech) is one of the most dominant forest species and makes a large evergreen tree. It can reach 50 m tall and live in excess of 500 years. The genus is not represented in the flora of the northern hemisphere but occurs in Chile (see above). *Eucryphia lucida* (leatherwood tree) is a Tasmanian endemic, with its attractive flowers forming the basis of the island's honey industry. *Atherosperma moschatum* (sassafras) is another large tree of the rain forest, with its timber used in the production of high-quality furniture. It often grows together with *Nothofagus cunninghamii*.

Phyllocladus aspleniifolius (celery top pine) can grow to be 30 m tall and live to 800 years old. Its common name reflects the fact that its leaves (actually flattened stems) look like those of the celery plant. Other trees include *Ceratopetalum apetalum* (coachwood), *Doryphora sassafras*, *Acmena smithii*, *Eucryphia moorei*, and *Nothofagus moorei* (180, 181).

Lagarostrobos franklinii (Huon pine) is a slow-growing conifer of the TRF. Trees may eventually reach 40 m tall and attain ages of around 2,000 years. Large natural stands occur in a designated world heritage area. *Athrotaxis selaginoides* (King Billy pine) and *A. cupressoides* (pencil pine, 182) are both large coniferous trees which grow at elevations above 600 m above sea level and can reach ages in excess of 1,200 years.

179 Details of the foliage and 'fruits' of the TRF conifer *Podocarpus nubigenus* from southern Chile. (Photo copyright of Martin Gardner.)

180 Foliage canopy of *Nothofagus moorei* (southern beech) growing in Australia.

181 Foliage of *Nothofagus moorei* from the TRF of western Tasmania.

182 Detail of the foliage of the conifer *Athrotaxis cupressoides* from the Tasmanian TRF.

CHAPTER 4
Temperate mixed evergreen forests
Stephen D Hopper, Erika Pignatti Wikus, and Sandro Pignatti

INTRODUCTION

This chapter focuses on temperate evergreen trees, especially those experiencing a Mediterranean climate. These, in part, have their origins in rainforest lineages from past periods of wetter climate (Cowling *et al.*, 1996; Dallman, 1998). Temperate evergreen forests (TEFs) occur in both the northern and southern hemispheres, and include the remarkably diverse acacias and eucalypts of Australia – in which the world's tallest hardwood tree species occur; the rich evergreen oak forests of the Mediterranean; the very striking southern beech forests of Chile, Tasmania, and New Zealand; and the evergreen temperate coniferous rain forests of California. The last of these has already been considered in

Chapters 2–3 and will not be further discussed here.

Among the more unusual trees found in the TEFs are a number of arborescent monocotyledon species, which include the South African *Aloe dichotoma* (quiver tree, **183**), the Californian *Yucca brevifolia* (Joshua tree, **184**), the Mediterranean *Dracaena draco* (dragon tree, **185**), the Australian *Dasypogon hookeri* (pineapple bush, **186**), and species of *Xanthorrhoea* (grass trees, **187**). Southwest Australia also has the endemic *Nuytsia floribunda* (Christmas tree, **188**), the only arborescent species of mistletoe. Dwarf forests include the stunning vertebrate-pollinated banksias and proteas of the southern continents, while woody irids form elfin forests in the fynbos of South Africa.

183 A lone *Aloe dichotoma* (quiver tree) stands erect among otherwise dwarf succulent karoo vegetation on the slopes of the Kamiesberg, Namaqualand, South Africa. (Photo copyright of Stephen D Hopper.)

184 An evergreen woodland of *Yucca brevifolia* (Joshua tree) among granite hills on the Barber Dam Loop Trail, Joshua Tree National Park, Mohave Desert, USA. (Photo copyright of Stephen D Hopper.)

185 Part of the crown of *Dracaena draco* (dragon tree) − note its long sword-shaped leaves, which are typical of monocotyledons. (Photo copyright of Bryan Bowes.)

186 Several specimens of *Dasypogon hookeri* (pineapple bush) growing up to 5 m tall in a forest at Leeuwin-Naturaliste National Park, southwest Australia. This monocotyledon species represents an Australian endemic order and family. The dasypogons are flanked by *Allocasuarina fraseriana* (common she-oak), with *Banksia grandis* behind. (Photo copyright of Stephen D Hopper.)

187 Several tall specimens of *Xanthorrhoea glauca* (grass tree) growing in Lamington National Park, southern Queensland, Australia. (Photo copyright of Bryan Bowes.)

188 A flowering *Nuytsia floribunda* (Christmas tree), the world's largest arborescent mistletoe, growing among early summer vegetation on the Swan Coastal Plain, Bunbury, southwest Australia. On the right, *Melaleuca preissiana* (modong) is visible. (Photo copyright of Stephen D Hopper.)

TEMPERATE EVERGREEN FORESTS OF AUSTRALIA

EUCALYPTUS (EUCALYPTS)

Although known from fossils in Argentina and New Zealand, eucalypts today are almost all endemic to Australia, but with a few species occurring on adjacent tropical islands (Brooker and Kleinig, 2001). There are more than 900 eucalypt species, and new ones continue to be described. Recent DNA sequence studies show that the eucalypts comprise four genera – the monotypic *Arillastrum* (from New Caledonia), *Angophora*, *Corymbia*, and *Eucalyptus*, in which most species still remain (Steane *et al.*, 2002).

Primarily due to their capacity to re-sprout or germinate prolifically after fire, eucalypts dominate Australian landscapes. On south coastal areas exposed to the fierce winds of the roaring forties (**189**), eucalypts form ground-hugging shrubs only 20 cm tall. Other species soar more than 90 m as lofty, straight-trunked trees. These include *E. regnans* (mountain ash, **190**) and *E. globulus* (Tasmanian blue gum) in Victoria and Tasmania; and *E. diversicolor* (karri) in the southwest of Western Australia. Australia's most widespread eucalypt is *E. camaldulensis* (river red gum), while *E. pauciflora* (snow gum) is a feature of high country. The multi-stemmed mallees (small eucalypts with large, burl-like lignotubers) include hundreds of *Eucalyptus* species growing in semi-arid environments (**191**).

Southwest Australia is one of two hotspots of eucalypt diversity (the other centred on Sydney, is not considered further in this chapter), and is the greatest centre of local endemism. This region is an ancient glaciated plateau, which shows a subtle complexity of soil and topographic variation. Up to 400 or more *Eucalyptus* species coexist here, responding to minor local environmental variations. Most are mallees, of which some species extend into low-rainfall areas, their deep roots tapping groundwater and enabling the formation of woodlands under desert-like conditions (**192**).

189 Wind-pruned mallees of *Eucalyptus diversifolia*, which are only 20 cm tall, growing as a dwarf heath on South Australia's Kangaroo Island.(Photo copyright of Stephen D Hopper.)

190 Trunk of *Eucalyptus regnans* (mountain ash). This specimen ('the big tree') is the world's tallest broadleaf tree and soars to 100 m at the Huon River, Tasmania. (Photo copyright of Stephen D Hopper.)

(karri) are an inspiring sight. The giant trees of three other species of *Eucalyptus* (tingles) are less well known. *E. guilfoylei* is the most widespread, *E. jacksonii* is less common, and *E. brevistylis* is rarer. All are extremely narrow endemics and are just able to persist in the highest rainfall margins of the south coast, in the Walpole-Denmark district.

Corymbia calophylla (marri) and *Eucalyptus marginata* (jarrah) are the other forest giants of the highest rainfall country. Marri belongs to a group of bloodwood eucalypts, the richly-coloured sap of which oozes from wounded trunks. It is of immense importance as a medicine tree to the Nyoongar Aboriginal people, and is also a major source of nectar, seeds, and insects for birds and arboreal mammals. *E. marginata* is the southwest's major timber species. It grows tallest on the massive laterite gravels of the Darling Range near Perth, but also grows high in the southern forests, intermixed with *E. diversicolor* and *Corymbia calophylla*.

Major understorey trees of these giant hardwood forests include *Allocasuarina decussata*, *Trymalium floribundum*, *Agonis flexuosa*, *Banksia grandis* (**194**), and *Acacia pentadenia*. In the family Proteaceae, there are more than 170 species of *Banksia* and *Dryandra*, which are mostly endemic to the kwongan (heath), woodlands, and forests of the southwest. Their fine roots help banksias obtain scarce nutrients from impoverished soils, while their large robust inflorescences attract birds, marsupials, and rodents as pollinators (**194**).

The tall forests occupy rich loams on valley slopes and well-drained flats. The margins of swamps in this highest rainfall country support *Agonis juniperina*, paperbarks such as *Melaleuca preissiana*, *Callistachys lanceolata* (native willow), and the highly distinctive grasstree *Kingia australis*. Occasionally, the waterlogged margins of creeks and rivers are home to *Eucalyptus patens*, *E. megacarpa*, and *Banksia seminuda* subsp. *seminuda*. On granite outcrops, dense low forests of many species, including *Eucalyptus cornuta*, *Melaleuca croxfordiae*, *Hakea elliptica*, *Banksia seminuda* subsp. *remanens*, and *B. verticillata*, occur. Some coastal consolidated dunes have forests of *Agonis flexuosa* (west Australian peppermint) and *Eucalyptus occidentalis* (mo or swamp yate), accompanied by understorey trees such as *Hakea*

oleifolia. Also conspicuous in this environment (but also present throughout the southwest) are the arborescent grass trees including *Xanthorrhoea platyphylla* and *X. preissii* (**195**).

194 An inflorescence of *Banksia grandis* (bull banksia) with *Tarsipes rostratus* (honey possum, an endemic mouse-sized southwest Australian marsupial) in search of its nectar. Millbrook Nature Reserve. (Photo copyright of Stephen D Hopper.)

195 A 4 m tall specimen of the arborescent cycad *Macrozamia fraseri*. The spike-like scapes of *Xanthorrhoea preissii* (balga grasstrees) are visible in the distance. Lake Indoon, southwest Australia. (Photo copyright of Stephen D Hopper.)

JARRAH AND PINEAPPLE BUSH FORESTS (*EUCALYPTUS MARGINATA* AND *DASYPOGON HOOKERI*)

Jarrah forests tend to have fewer tree species present than the highest rainfall tall forests. Most often, *Eucalyptus marginata*, *Corymbia calophylla*, and *Allocasuarina fraseriana* dominate the uplands and slopes, with *Eucalyptus patens*, *E. megacarpa*, *E. rudis*, and *Melaleuca* along the waterlines and swamps. Nevertheless, isolated pockets of unusual trees, such as *Eucalyptus lane-poolei*, *E. laeliae*, *E. wandoo*, *Corymbia haemotoxylon*, and *Allocasuarina huegeliana*, occur. Also, pockets of white sand support a low forest of *Banksia attenuata*, *B. menziesii*, and *B. ilicifolia*, with *B. littoralis* found in waterlogged sites.

In the low forest of the Leeuwin-Naturaliste National Park, scattered groves of the arborescent monocotyledon *Dasypogon hookeri* (pineapple bush, family Dasypogonaceae) occur. These columnar plants grow up to 5 m tall and are topped with pineapple-like foliage and drumsticks for flowers (**186**). The relationships of the family remain unelucidated, despite recent DNA studies, but it is certain that their ancestors date back 120 million years to the Cretaceous period.

ARBORESCENT CYCADS

Cycads are gymnosperms, the seeds of which are naked (as is also the case with other conifers), in contrast to flowering plants (angiosperms), the seeds of which develop within a fruit. Most extant Australian cycads form large palm-like rosettes but some, such as *Macrozamia fraseri* (**195**) in the southwest, and *Lepidozamia peroffskyana* (**196**) in southeast Queensland, develop a trunk and occasionally form groves of low forest 2–3 m high. The large female cones produce big red seeds (**197**) that are highly toxic but were eaten by Australian Aborigines after prolonged leaching in water. However, unwary early European explorers were often violently ill after they ate the seed raw. This was the plight of some of Vlamingh's men who first explored the Swan River, by Perth in 1697. The account by the expedition's Surgeon Torst of this incident is as follows:

'… they brought me the nut of certain fruit trees, resembling in form the 'droiens'.

Having the taste of our large Dutch bean; and those which were younger were like a walnut. I ate five or six of them and drank the water from a small pool; but, after an interval of about three hours, I and five others who had eaten these fruits began to vomit so violently that we were as dead men.' (Playford, 1998).

BEYOND THE TALL FORESTS

The Swan Coastal Plain supports less lofty forests and occasional low forest stands of *Kunzea ericifolia* and *Dryandra sessilis*, but perhaps the signal tree is *Eucalyptus gomphocephala*, which forms forests up to 40 m tall on the west coastal limestone soils from Busselton to Jurien Bay. On Garden Island, offshore near Perth, one of southwest Australia's few coniferous forests is located, with *Callitris preissii* forming dense-canopied verdant stands. The conifer woodland of *Callitris tuberculata* on Bald Island is estimated to be 150 years old and is richly endowed with quokkas, a small marsupial. Another striking

196 Two large trees of *Lepidozamia peroffskyana*; these were formerly part of a forest stand but are now lone survivors in the garden of a house at Mt Tambourine, southeast Queensland, Australia. These ancient specimens are about 2.5 m tall and the front specimen bears a large (old) female cone. (Photo copyright of Bryan Bowes.)

197 Massive female cone of *Lepidozamia peroffskyana*; this has been pecked open to reveal its bright red seeds, which are highly toxic for humans. (Photo copyright of Bryan Bowes.)

198 Flowers and buds of *Eucalyptus caesia* subspecies *magna* (caesia), illustrating the colourful eucalypts, which are vertebrate-pollinated, and are a feature of the southwest Australian vegetation. (Photo copyright of Stephen D Hopper.)

coastal tree on these islands is *Melaleuca lanceolata*.

The diversity of forest trees into the semi-arid wheatbelt and adjacent goldfield's woodlands is remarkable. *Eucalyptus salmonophloia* and *E. salubris* form dense low forests and dominate the relatively fertile clay loams of the broad valley floors. *Eucalyptus wandoo* occupies the slopes of the western wheatbelt, interdigitating with the drier margins of the *E. marginata* (jarrah) forest. *Eucalyptus capillosa* (inland wandoo) occurs in the eastern wheatbelt and goldfields, among decomposing granite breakaways. Various species of *Eucalyptus* (mallets), for example *E. gardneri*, *E. argyphea*, *E. clivicola*, and *E. astringens*, form low forests with little understorey on the flat-topped lateritic mesas. On granite outcrops, forests of *Allocasuarina huegeliana* (rock oak), *Acacia lasiocalyx*, *Hakea petiolaris*, and an understorey of eucalypts and other small trees occurs.

Southwest Australia has many species with large flowers, attractive to bird and mammalian pollinators (194, 198). *Eucalyptus macrocarpa* (mottlecah) has brilliant red flowers, 10 cm across. Its rare relative, *E. rhodantha* (rose mallee), is confined to just two small areas in the northern wheatbelt. *E. caesia* is rare and has a few-trunked mallee habit. During the winter months, its bright pink flowers attract honeyeaters in multitudes (198). More than 100 of the Australian eucalypts are rare or threatened, and most occur in temperate agricultural regions where the native vegetation has been extensively destroyed to make way for wheat and sheep.

MARLOCKS, MOORTS, AND MALLEES

Eucalypts of the semi-arid southwest have evolved a range of habits, so distinctive that Aboriginal people bestowed unique names on them. Marlocks are effuse small eucalypts, often forming dense low forests and thickets in the wheatbelt and on granite outcrops. They include *Eucalyptus conferruminata* (Bald Island marlock) and the moorts *E. platypus* and *E. utilis*. After a fire, such species carpet the ground with woody fruits that open after the heat of burning. As soon as rain moistens the ground, the seeds germinate and thrive on the nutrient-rich ashbed, so that within a few years a dense low forest develops again.

Multi-stemmed *Eucalyptus* species (mallees, 191) populate low forests over vast areas of semi-arid country (Pignatti Wikus *et al.*, 2001). There are more than 200 species of mallee in the region, representing one of the world's greatest radiations within a single genus of woody plants. Following the death of the aerial tree, mallees can re-grow from the sprouting underground lignotuber ('mallee root').

Another extraordinary southwest endemic is *Nuytsia floribunda* (Christmas tree), the spectacular orange flowers of which appear early in the southern summer. It is an arborescent member of the family Loranthaceae (**188**), with parasitic roots that run through the sandy soil to girdle roots of adjacent host species and tap them for water. Consequently, its foliage remains cool through the hot southwest Australian summer.

THE UNEXPLORED TREE DIVERSITY OF SOUTHWEST AUSTRALIA

The semi-arid low forests (**192**), and the mallee communities of southwest Australia (**191**), are havens of tree diversity. New species continue to be discovered as less-explored corners of the wheatbelt, goldfields, and mallee regions are investigated. Even the high-rainfall southern forests, which have long been exploited for timber, surprisingly continue to reveal previously unknown trees. *Eucalyptus virginea*, described in 2004, grows up to 20 m tall and 1 m in diameter, but is confined to a few hectares on the granite slopes of Mt Lindesay – its entire global habitat. That such a tall handsome tree could remain undetected by botanists for so long shows how much more there is to learn about these forests.

TEMPERATE EVERGREEN FORESTS OF TASMANIA AND NEW ZEALAND

There are many interesting trees in the cool temperate forests of Tasmania and New Zealand (**199, 200**, Kirkpatrick and Backhouse, 1985; Salmon, 1986; Dawson and Lucas, 2000). Several species of *Nothofagus* (southern beeches) are often dominant, as are conifers such as *Libocedrus bidwillii*, *Podocarpus*, and various araucarian species. The tree fern *Dicksonia antarctica* is common and grows 2–3 m high on an unbranched trunk (**201**). In Tasmania, an unusual-looking tree is *Richea pandanifolia* (pandani, **199**), belonging to the family Ericaceae but resembling the unrelated southwest Australian dasypogons (**186**) in habit. *R. pandanifolia* is confined to the wetter western side of Tasmania. It is common in rainforest and conifer forests but is also present in the alpine zone. The New Zealand palm *Rhopalostylis sapida* (nikau) is the world's southernmost member of the palm family; it attains 10 m in height and is found in the lowland forests of both islands (**200**).

199 Cool temperate coniferous forest in Tasmania, with a single specimen of *Richea pandanifolia* (pandani, Ericaceae) in the mid-centre. It is flanked by *Athrotaxis cupressoides* (pencil pine) and *A. selaginoides* (King William pine), while *Microstrobus niphophilus* is in the foreground. Lake Dobson, Mt Field National Park. (Photo copyright of Stephen D Hopper.)

200 New Zealand's temperate mixed evergreen forest comprises podocarp conifers and many genera of angiosperms, as well as tree ferns. *Rhopalostylis sapida* (nikau), the world's southernmost palm, is seen here growing 10 m high on the lower slopes of the hill. (Photo copyright of Stephen D Hopper.)

TEMPERATE EVERGREEN FORESTS OF SOUTH AFRICA

Forests are relatively rare in the Greater Cape region of South Africa (Cowling *et al.*, 1997; van der Merwe, 1998; Pauw and Johnson, 1999; Goldblatt and Manning, 2000). However, pockets of afro-montane forest up to 30 m tall are found in well-watered sites sheltered from fire, especially in the Knysna–Tsitsikamma area on the south coast. Here occur various species typical of the forests further east and north, including *Afrocarpus falcatus* (Outeniqua yellow-wood, 202), *Olinea ventosa* (hard pear), *Ocotea bullata* (stinkwood), *Rapanea melanophloeos* (Cape beech), *Curtisia dentata* (assegai), *Olea capensis* subspecies *macrocarpa* (ironwood), *Apodytos dimidiata* (white pear), and *Cunonia capensis* (red alder, 203). Various arborescent species such as *Halleria lucida* (tree fuchsia) and *Cyathea capensis* (tree fern) form an understorey. To the west, the number of tree species diminishes, whereas the forests further east and northeast are richer in variety.

201 A specimen of the tree fern *Dicksonia antarctica*, with an unbranched trunk which terminates in a spreading cluster of fronds (leaves) 2 m or more long. (Photo copyright of Bryan Bowes.)

202 This tall specimen is an example of the evergreen conifer *Afrocarpus falcatus*, which was growing by the Storms River, South Africa. (Photo copyright of Stephen D Hopper.)

203 *Cunonia capensis* (red alder) growing on the steep upper slopes of Table Mountain, South Africa. (Photo copyright of Stephen D Hopper.)

On the Tsitsikamma coast, specimens of *Afrocarpus falcatus* (**202**), although slow-growing, may be 800 years old and reach heights of 45 m. Their hard-shelled seeds are dispersed mainly by Egyptian fruit bats, Knysna louries, vervet monkeys, and chacma baboons. Close to the coast, the dry scrub forest grows 6–12 m high and is dominated by *Sideroxylon inerme* (milkwood), *Pterocelastrus tricuspidatus* (candlewood), and *Cassine peragua* (Cape saffron). In places, this scrub forest merges with coastal thickets of *C. maritima* (dune saffronwood), *Olea exasperata* (coast olive), and *Euclea racemosa* (sea guarri). Throughout the Cape region, rivers and creeks often have low forest of *Brabejum stellatifolium* (wild almond), *Ilex mitis* (waterboom), and *Metrosideros angustifolia* (smalblad).

A low, closed forest occurs on granite outcrops east of Cape Town (**204**). It contains species such as *Maytenus acuminata* (sybas), *Maytenus oleoides* (klipkershout), *Kiggelaria africana* (wild peach), *Olea europea* (wild olive), *Podocarpos latifolius* (yellow-wood), and *Cassine schinoides*. Red-winged starlings frequently fly into these forest pockets and, after eating the fleshy fruits, distribute the seeds widely. The seeds of *Cassine peragua* and *Olea capensis* (Cape saffron and ironwood) germinate more rapidly after passage through the gut of the local Rameron pigeons.

Outside these forests, trees of tall stature are rare.

204 Afro-montane forest evades fire by clinging to the steep slopes of Gordon's Peak on the Paarl granite massif east of Cape Town, South Africa. (Photo copyright of Stephen D Hopper.)

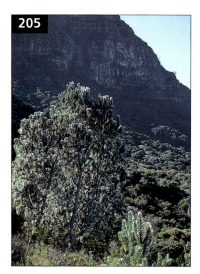

205 This specimen of *Leucadendron argenteum* (silver tree) exemplifies one of the few fynbos species that grow as trees in South Africa's Cape Region. Table Mountain, Kirstenbosch. (Photo copyright of Stephen D Hopper.)

206 Small trees, growing up to 4 m tall, of *Protea nitida* (waboom, left) and *Heerea argentea* (kliphout) on Paarlberg, east of Cape Town, South Africa. (Photo copyright of Stephen D Hopper.)

207 *Widdringtonia nodiflora* (mountain cypress) is one of the few native conifers of the Cape Region, near Dassiehoek, South Africa. (Photo copyright of Stephen D Hopper.)

A famous endemic confined to the Cape granite soils is *Leucadendron argenteum* (silver tree, **205**), which is widely cultivated for its silky leaves. A few other plants on these granite soils, such as *Protea nitida* (waboom), *Heerea argentea* (kliphout), and *Widdringtonia nodiflora* (**206, 207**), form small trees.

The fynbos on siliceous sands has fewer trees, and small shrubs dominate instead (**208**). These are mainly members of the families Proteaceae and Ericaceae, graminoids (Restionaceae, Cyperaceae), and geophytes (Iridaceae, Amaryllidaceae, Hyacinthaceae). However, a woody habit is found in certain genera of the Iridaceae (Goldblatt, 1993); in well-watered sites these can form elfin forests 2–3 m tall (**209**). Some legumes also form small

patches of forest, for example *Virgilia oroboides* (keurboom) growing to 20 m, and *Psoralea pinnata* and *P. fleta* growing up to 7 m tall. *Widdringtonia cedarbergensis* (Clanwilliam cedar) also forms groves up to 8 m tall in the Cederberg mountains.

The succulent karoo biome, ranging northwards from near Cape Town along the west coast and into southern Namibia, is similarly dominated by low shrubs and geophytes. However, *Aloe dichotoma* (quiver tree) is a distinctive emergent, sometimes attaining 5–6 m in height (**183**). The large yellow flowers provide abundant nectar for sunbirds and baboons in winter. *Acacia karoo* (sweet thorn) forms low forests 5–6 m tall along water courses (**210**). Species of *Ficus* (**211**) and

208 Typical appearance of fynbos with outcrops of Table Mountain sandstone. The emergent shrubs are *Leucospermum conocarpodendrum* (Proteaceae). Cape of Good Hope Nature Reserve, South Africa. (Photo copyright of Stephen D Hopper.)

209 A specimen of *Witsenia maura* (bokmakieriestert), an unusually woody member of the Iridaceae, which forms elfin forest in Cape fynbos. Cape Point Nature Reserve, South Africa. (Photo copyright of Stephen D Hopper.)

210 *Acacia karoo* (sweet thorn) on the Wilgerhoutsrivier beneath Stalberg, east of Garies, Kamiesberg, Namaqualand. (Photo copyright of Stephen D Hopper.)

211 *Ficus ilicina* (rock fig) on a granite outcrop east of Honderklipbaai, Namaqualand. (Photo copyright of Stephen D Hopper.)

213 *Erythrophysa alata*, a summer-deciduous and bird-pollinated shrub growing on a granite outcrop, Kamieskroon, Namaqualand. (Photo copyright of Stephen D Hopper.)

212 *Dodonaea viscosa* (sand olive) on a granite outcrop, Kamieskroon, Namaqualand. (Photo copyright of Stephen D Hopper.)

shrubs such as *Dodonaea viscosa* (**212**) and the summer-deciduous *Erythrophysa alata* (**213**) occur on rock outcrops.

TEMPERATE EVERGREEN FORESTS OF NORTHERN AND CENTRAL CHILE

Chile extends 4,200 km north–south but only 160 km east–west. It is bordered on the east by the high Andes and on the west by prominent coastal ranges up to 2,200 m high, with the Valle Central lying between them. The climate ranges from severe desert in the north, through Mediterranean in central Chile, to colder in the south and glacial near Tierra del Fuego. Its forests and woodlands have an interesting diversity of trees (Rottmann, 1988; Hoffmann, 1994, 1995; Zegers, 1995; Dallman, 1998).

The rosaceous *Polylepis* (quenoa) forms low forests at 3,800–4,500 m elevation in the northern Chilean Andes, the highest altitude reached by broadleaf trees. The twisted trunks of quenoa bear a brownish multi-layered papery bark, which helps to insulate them against extreme weather. Further south, the conifer *Austrocedrus chilensis* (mountain cypress, **214**) occupies similar high elevations.

Some specimens of *Jubaea chilensis* (Chilean palm), reaching up to 23 m in height, occur in the evergreen forest of La Campana National Park (**215**). The palm does not flower until 30 years old but then bears edible fruits (coquitos). The trees are often tapped destructively for their sweet sap.

Near Copiapo, the riverbanks support a low forest of *Geoffroya decorticans* (chañar, **216**), *Acacia caven* (espino), *Prosopis chilensis* (algarrobo), *Schinus polygamus* (huingan), and *Schinus molle* (pepper tree). An unusual northern outlier of the southern forests occurs on the crest of mountains in the high humidity of Fray Jorge National Park; here the trunks of *Drimys winteri* (canelo, **217**) and *Aetoxicon punctatum* (olivillo) are festooned with moss and vines, although surrounded by arid shrublands.

Much of the former forest of central Chile is now agricultural but it has been planted sporadically with various eucalypts and *Pinus radiata* (Monterey pine). However, the native evergreen trees include *Quillaja saponaria* (quillay, **218**), *Peumus boldus* (boldo, which is aromatic and with edible fruit), *Lithrea caustica* (litre, **219**),

214 Large specimen of *Austrocedrus chilensis* (mountain cypress). (Photo copyright of Martin Gardner.)

215 Several Chilean palms (*Jubaea chilensis*), which may attain 23 m at maturity. (Photo copyright of Martin Gardner.)

216 Low forest of *Geoffroya decorticans* (chañar). (Photo copyright of Martin Gardner.)

217 Flowers of *Drimys winteri* (canelo). (Photo copyright of Martin Gardner.)

218 Specimen of *Quillaja saponaria* (quillay). (Photo copyright of Martin Gardner.)

219 Fruits of the evergreen *Lithrea caustica* (litre). (Photo copyright of Martin Gardner.)

82

220 Stand of evergreen *Maytenus boaria* (maiten). (Photo copyright of Martin Gardner.)

221 Stripped bark from *Quillaja saponaria* (*cf.* 218), from which a natural detergent is extracted. (Photo copyright of Martin Gardner.)

222 Specimen of *Cryptocarya alba* (peumo). (Photo copyright of Martin Gardner.)

223 Branches of *Luma apiculata* (arrayan). (Photo copyright of Martin Gardner.)

Maytenus boaria (maiten, 220), the deep-rooted *Prosopsis chilensis* (algarrobo), and *Acacia caven* (espino). Specimens of *Quillaja saponaria* are highly sought after for their bark (221) which contains a natural detergent. In regions of moister soils, 10–20 m tall forests of *Cryptocarya alba* (peumo, 222), *Maytenus boaria* (maiten), *Peumus boldus*, and *Beilschmiedia miersii* (belloto) occur. Regions of waterlogged swamps and riverbanks support the growth of *Salix chilensis* (Chilean willow), *Myrceugenia exsucca* (petra), *Luma apiculata* (arrayan, 223), *Drimys winteri* (217), and *Crinodendron patagua* (patagua).

On the coastal slopes to the west of Santiago grow evergreen forests of *Maytenus boaria* (220), *Peumus boldus*, and *Quillaja saponaria* (218), while the mountain peaks are dominated by the deciduous *Nothofagus obliqua* (roble). Further south, *Beilschmiedia miersii*, *Persea lingue* (lingue), and *Aetoxicon punctatum* grace the coastal hills, some of the latter species attaining 25 m in height.

TEMPERATE MIXED EVERGREEN FORESTS OF THE MEDITERRANEAN

This zone includes the southern parts of Europe, North Africa, and a small portion of western Asia. The evergreen forests are confined to the warmer areas of these regions, whereas in colder areas and on the mountains deciduous forest occurs. The evergreen trees consist of a mixture of broadleaf species (mainly from the genus *Quercus*, oaks, 224–229) and conifers such as species of *Pinus* (pine),

224 The evergreen forest of Orgosolo lying at about 700 m in the mountains of central Sardinia. *Quercus ilex* forest is well developed here but, in the shallow depression with deeper soil, pasture is now present.

225 Foliage of *Quercus ilex* – note its yellow-green, glossy evergreen leaves. (Photo copyright of Bryan Bowes.)

226 On the driest and warmest slopes in Sardinia, the dark-green *Quercus ilex* forest is replaced by *Juniperus oxycedrus* and *J. phoenicea*. The juniper scrub is recognized by its light green colour.

227 In Spain and Portugal the evergreen *Quercus ilex* is replaced by *Q. rotundifolia*.

228 The trunk of an old *Quercus suber* (cork oak) is covered by its bark. In former times the harvesting of cork bark was very lucrative, but now it is nearly abandoned in Italy although still gathered in Portugal and Spain (**229**).

229 *Quercus suber* soon after harvesting the cork; the reddish phellogenous layer will regenerate a new cork crop in 10–15 years.

230

230 *Juniperus phoenicea* growing as a small tree in the coastal formation of Sardinia, which is the remains of a former, much more common, belt of vegetation.

231

231 The North African woodland with *Ceratonia siliqua*. This tree also occurs rarely in protected biotopes along the southern European coast.

Abies (fir), *Cedrus* (cedar), *Juniperus* (juniper, **226**, **230**), and *Cupressus* (cypress). The most widespread evergreen oak (*Quercus ilex*, **224**, **225**) has a striking similarity to the North American evergreen oaks (*Quercus virginiana*, *Q. chrysolepis*).

The distribution of the arborescent flora depends mainly on the climate. The broadleaf trees are mostly restricted to areas with a true Mediterranean climate in which the annual average temperature shows little variation (from 14–18°C), but there are between two and five months of summer drought. The conifers also occur in these same localities, but some species can extend into the colder climates of mountains; for example, *Cedrus atlantica* grows up to altitudes of 2,000 m in North Africa. Palms are mostly absent in the Mediterranean. The only native species (*Chamaerops humilis*) hardly reaches 2 m and generally does not occur in forest communities.

In a broad overview the most striking features of the Mediterranean forests are as follows:

- *Quercus ilex* (holm oak) forest (**224**, **225**). This forms a very dense evergreen community where the tree layer is mainly composed of *Q. ilex* (**225**, Braun-Blanquet, 1936), a dense layer of shrubs and climbers, and an impoverished herb layer reduced to few species, probably because of insufficient light. Many components of this forest have affinity with the tropical flora, for example *Pistacia* (pistachio), *Smilax*, *Myrtus* (myrtle), and the family *Oleaceae*. There are three distinct subspecies of *Quercus ilex*, which occur in Spain, the North African Maghreb, and southern Europe

(in France, Italy, Greece, and many islands). Towards the region of the eastern Mediterranean, this vegetation becomes progressively rarer.

- *Quercus suber* (cork oak) forest (**228**, **229**). This is mostly a relatively open evergreen community in which oaks and pines grow with an understorey of *Cistus* scrub. In general, this forest occurs as a permanently disturbed community as the consequence of repeated fires, and the soils are relatively leached. It is mainly distributed in the western Mediterranean region and covers large areas of the Iberian Peninsula and Morocco. It is rather rare in southern Italy and Sicily, and completely lacking in the eastern Mediterranean.
- Other oak communities. These are characterized by evergreen species such as *Quercus coccifera* and *Q. calliprinus* (it remains to be demonstrated whether these are really distinct species), or by semi-evergreen species, such as *Quercus troiana* and *Q. ithaburensis*, in the eastern Mediterranean belt.
- Open scrub vegetation with evergreen elements. In the warmer and drier parts of the Mediterranean, species such as *Ceratonia siliqua* (**231**), *Olea oleaster*, *Pistacia lentiscus*, and *Euphorbia dendroides* predominate, while evergreen oaks are limited to wetter or more montane habitats. This vegetation is widespread in North Africa but occurs only in the hottest regions of the southern coasts of Europe.
- Mediterranean pine forests. *Pinus halepensis* (Aleppo pine), *P. pinaster* (maritime pine), and *P. pinea* (stone pine) form open communities. In

coastal areas or at lower altitudes an understorey of endemic low shrubs occurs, and the latter are often aromatic and thorny.

- Mediterranean mountain pine forests. Several pine species occur only in mountain habitats. These include *Pinus leucodermis* (= *P. heldreichii*) in the southern parts of Italy and the Balkans, and several subspecies of the *Pinus nigra* (Austrian pine) complex (subsp. *laricio* in Corsica, southern Italy, and Sicily, subsp. *pallasiana* in the Balkan Peninsula, subsp. *salzmannii* in France and Spain). Several species of *Abies* (fir) also occur as restricted endemics, such as *A. pinsapo* (Spanish fir) in southern Spain, *A. nebrodensis* in Sicily, and *A. cephalonica* (Grecian fir) in Greece. The forests of *Pinus brutia* in the eastern Mediterranean, together with the *Cedrus* (cedar) forests in Lebanon, adjacent areas of North Africa, and western Asia, should also be included in this grouping.

ORIGINS OF MEDITERRANEAN MIXED EVERGREEN FORESTS

The complex taxonomic relationships described above, and the plant distribution patterns, can only be understood by taking into account the effects of both the past and the present (Pignatti, 1998). At the end of the Miocene epoch (*ca.* 5.5 mya) and into the succeeding Pliocene period, the Mediterranean region had sub-tropical vegetation, which mainly consisted of evergreen species with an affinity with the east Asian and North American flora. During the earlier Messinian period (about 7–6 mya), the Gibraltar Strait had emerged and the Mediterranean became separated from the oceans and dried up. Most of the region became a salty desert, similar to the present-day Dead Sea, and the forest vegetation survived only in mountain habitats where *Pinus*, *Abies*, and some broadleaved evergreen elements, such as species of *Ilex* (holly, **232**) and *Taxus* (yew), populated refugial areas. This flora was derived from the evergreen temperate flora of the previous periods.

Later, after the Messinian period, the Mediterranean became newly filled to form an inland sea, and vegetation of the present-day type spread everywhere. However, this situation subsequently changed as a consequence of the cold periods during the Quaternary, from *ca.* 1.8 mya onwards. In the Mediterranean there was no real glaciation, but low temperatures in mountain habitats had severe consequences on their evergreen flora with tropical affinities. All over the Mediterranean basin, many species became extinct while others, such as *Zelkova* (wing nuts) and *Abies nebrodensis*, were reduced to small populations. After the cold periods, the evergreen elements had little capacity to re-colonize the surrounding areas. Instead, a relatively new and vigorous flora took over, comprising deciduous temperate trees such as *Fagus* (beech), *Acer* (maple, sycamore), *Alnus* (alder), and deciduous species of *Quercus* (Quézel and Medail, 2003). However, this deciduous vegetation also absorbed some evergreen elements of the previous temperate flora, such as *Ilex aquifolium* (**232**, **233**).

232 In the foreground is a specimen of *Ilex aquifolium* (holly) growing on the western coast of Sardinia. This evergreen shrub/tree is growing together with the deciduous oak *Quercus pubescens* (seen in the background).

233 Foliage of *Ilex aquifolium* – note its dark green, glossy leaves with their sharp marginal prickles. (Photo copyright of Bryan Bowes.)

HUMAN IMPACT ON MEDITERRANEAN EVERGREEN FORESTS

Under present-day conditions some evergreen forests can be regarded as rare climax communities. These include *Quercus ilex* forest, *Olea-Ceratonia* scrub, *Cedrus*, and other formations. However, the Mediterranean area has been populated by man for a very long time and most vegetation has been influenced by human management. The first unambiguous evidence of the use of fire is dated at 420,000 years BP, and fire has been largely used for hunting and clearing the landscape. *Homo erectus*, later *Homo sapiens* (modern man), were present in the whole area during the cold periods of the Quaternary (from *ca.* 1.8 mya to the present day). In prehistorical times (10–20,000 years BP) goats, sheep, and cattle were domesticated; consequently, the Mediterranean vegetation was heavily impacted by grazing and fire.

The present plant cover of the Mediterranean is the consequence of agricultural land management (Di Castri *et al.*, 1981). With the neolithic revolution (after 9000 BP), agriculture spread over the whole Mediterranean basin, and the natural forest vegetation was drastically reduced. This management transformed the soil and landscape, and it is very likely that the flora evolved in the niches opened by human impact. The forest belt was transformed mainly into a dense evergreen scrub (macchia, maquis) with a floristic composition similar to the forest but only 3–5 m high. The macchia is often maintained by the coppicing of arborescent species, which avoids the regrowth of the tree layer. With greater human impact, the vegetation has been reduced to communities of low shrubs and annual herbs – the garrigue. In the present day, examples of climax vegetation occur extremely rarely (if at all) and large areas have become irreversibly desertified.

The structure of the Mediterranean vegetation was an important factor for the development of its civilization, culture, and social life, but conversely human activity played a fundamental role in the shaping of the plant cover in the area. Indeed, the transmission of Mediterranean land usage practices by western cultural technology has had a profound impact on all the Earth's temperate mixed evergreen forests.

CHAPTER 5

Tropical and sub-tropical rain and dry forests

Ghillean T Prance

INTRODUCTION

Tropical moist forests are estimated to cover an area of 11.7 million square kilometres of the humid tropics, where there is abundant rainfall and little seasonality. Tropical dry forests occur in areas of the tropics with seasonal drought caused by a long dry season, and cover about 2.5 million km² (Groombridge and Jenkins, 2002).

The area covered by tropical rain forests and tropical dry forests is actually a mosaic of many different types of forest. The causes for this variation are many, linked to soil types, climate, topography, and history, but the distribution of individual genera and species of trees is often crucial for the definition of the various types of forest. For example, the Amazon rain forest region is often defined by the distribution of the rubber genus *Hevea*. Within this broad category of the Amazon rain forest there are many different types of forest (Pires and Prance, 1985; Prance, 1989), such as upland non-flooded forest (terra firme, **234–237**), periodically inundated várzea and igapó forests (**238–243**; see also Prance, 1979), and coastal mangrove forests (**244–246**).

The species diversity of forests varies from place to place, with parts of western Amazonia and the Atlantic coastal forest of Brazil being the most species diverse (*Table 2*). The areas with wetter and less seasonal climates tend to be the most species diverse, and evergreen rain forest is more diverse than seasonal semi-deciduous forest types. Tropical rain forests contain a great diversity of trees, and up to 300 plus species of trees of 10 cm diameter and above may be present per hectare of forest. A few of the most important species that occur in the different types of tropical forest around the world will be highlighted in this chapter.

234 Emergent tree of *Cariniana estrellensis* in the Atlantic coastal rain forest of Brazil. Less than 7% of the original amount of this type of forest remains.

235 Brazil nut fruit at the top of a tree of *Bertholletia excelsa*, a typical terra firme forest tree of Amazonia.

236 Buttressed trunk of a species of fig (*Ficus*) in the upland forest of French Guiana.

237 Trunks of upland trees in French Guiana with several lianas (woody vines), which are common in tropical forests.

238 *Couratari tenuicarpa* (Lecythidaceae), a typical tree of the floodplain várzea forest.

239 *C. tenuicarpa* flowers in a leafless condition giving a show to attract its pollinators.

240 Flowers of *C. tenuicarpa* and their bee pollinators.

241 Pollen laden bee after visiting flowers of *C. tenuicarpa*.

242 Species of *Ficus* (fig) growing in the floodplain forests of Amazonia where the river level changes as much as 12 m between the wet and the dry seasons.

243 *Triplaris surinamensis* is an ant-inhabited tree typical of the Amazonian floodplain várzea and igapó forests.

244 Mass of stilt roots of the red mangrove (*Rhizophora*), the most common and characteristic tree of coastal mangrove forests. (Photo taken in Cameroon.)

245 Pneumatophores (breathing roots) are a common feature of mangrove forest trees. (Photo taken in Bunaken, Indonesia.)

246 *Bruguiera gymnorhiza* (black mangrove) showing flowers, the fruits of which later bear viviparous seedlings. (Photo copyright of Bryan Bowes.)

Table 2 Tree species diversity of some tropical rain forests on non-flooded ground, showing number of species, of 10 cm or more in diameter, per hectare

Number of species per hectare	Locality	Reference
311	Allpahuayo, Lor., Peru	Vásquez-Martínez and Phillips (2000)
300	Yanomono, Lor., Peru	Gentry (1988)
289	Mishana, Lor., Peru	Gentry (1988)
271	Rio Juruá, Am., Brazil	da Silva et al. (1992)
270	Serra Grande, Bahia, Brazil	Thomas and Carvalho (1993)
244	Añangu, Ecuador	Balslev et al. (1987)
228	Papua New Guinea	Wright et al. (1997)
228	Yasuni, Ecuador	Balslev et al. (1987)
227	Jenaro Herrera, Lor., Peru	Spichiger et al. (1996)
223	Mulu, Sarawak	Proctor et al. (1983)
196	Caxiuanã, Pa., Brazil	Almeida et al. (1993)
188	Camaipi, Ap., Brazil	Mori et al. (1989)
181	Tambopata, Peru	Gentry (1988)
173	Mocambo, Pa., Brazil	Cain et al. (1956)
162	Rio Xingu, Pa., Brazil	Campbell et al. (1986)
147	Caxiuanã, Pa., Brazil	Almeida et al. (1993)
136	Jaru, Ro., Brazil	Absy et al. (1988)
131	Oveng, Gabon	Reitsma (1988)
118	Rio Xingu, Pa., Brazil	Campbell et al. (1986)
103	Machadinho, Ro., Brazil	Absy et al. (1988)
81	Alto Ivon, Bolivia	Boom (1986)

THE FORESTS OF TROPICAL AMERICA

The rain forests of the Americas extend from southern Mexico to southern Brazil and the northernmost corner of Misiones Province in Argentina. The main blocks of rain forest occur in the Amazon region, the Atlantic coast of Brazil (**234**), the Pacific coast of Colombia and northern Ecuador, and into Central America in parts of Panama, Costa Rica, and Mexico. The rain forest area is interspersed with cloud forest (**247**, **248**), drier semi-deciduous forests, and savannas depending mainly on local rainfall quantity and seasonality (**249**). The dominant plant families vary slightly from region to region (*Table 3*), but the families Leguminosae, Sapotaceae, Lecythidaceae, and Arecaceae are the most important in Amazonia, and the Myrtaceae in the Atlantic coastal forest.

NEOTROPICAL FOREST ON TERRA FIRME

This is the most extensive forest type in Amazonia, where it occupies about 51% of the region. Important areas also occur in Central America, the Chocó of Colombia, and the Atlantic coastal region of Brazil. Neotropical forest occurs in areas of high rainfall and without a long dry season, on the plateaux above the flood level of the rivers. The forest is characterized by a closed canopy with some emergent species, and its height is usually about 25–35 m with emergents growing up to 50 m. The species diversity of trees varies from about 80 to 310 per hectare depending on soil and climatic factors (*Table 2*). Typical emergent trees are described below.

Bertholletia excelsa Humb. and Bonpl

The Brazil nut tree (**235**, family Lecythidaceae) is most characteristic of the non-flooded forests of

Amazonian Brazil, Peru, and Bolivia. Its distribution has probably been extended by indigenous populations because of its usefulness as a source of food and oil. It is one of the largest trees of the terra firme forest, often attaining a height of 50 m and a trunk diameter of over 2 m. It flowers in November/December at the onset of the rainy season, and the woody fruit with seeds (the nuts of commerce) take 14 months to mature. Each fruit (235) contains 14–24 nuts and the fruits fall to the ground in January/February. They are either harvested by human nut gatherers or extracted by agoutis, their natural agent of seed dispersal.

Dinizia excelsa Ducke

This leguminous tree is well named excelsa (or large) because it is the biggest tree in the terra firme forests of Amazonia, growing in the same habitat as the Brazil nut. It often reaches 55 m in height by up to 3 m in diameter at the base, and emerges far above the forest canopy. Some other typical trees of the Amazon forest on terra firme include *Caryocar villosum*, *C. glabrum* (Caryocaraceae), *Goupia glabra* (Celastraceae), many palms such as *Oenocarpus bacaba* (250) and *Astrocaryum aculeatum*, and many other Leguminosae such as *Parkia pendula*.

247 View of submontane forest at Auaris, Brazil, near the Venezuelan frontier in the Parima Mountains.

248 Granite outcrop of Serra Curicuriari in the upper Rio Negro region of Brazil. The hill attracts cloud, and even at a relatively low altitude (700 m) it is covered by dwarf cloud forest.

249 Cerradão or savanna forest typical of the more seasonal drier areas of central Brazil.

250 *Oenocarpus bacaba* (bacaba palm), a characteristic species of the palm forests of Amazonia. The fruits contain an edible pulp around the seeds.

Table 3 Dominant plant families from selected tropical inventories

A. NEOTROPICS

I   JENARO HERRERA, LORETO, PERU
    (Spichiger *et al.*, 1996)
    1. Sapotaceae
    2. Leguminosae
    3. Moraceae
    4. Chrysobalanaceae

II ALLPAHUAYO, LORETO, PERU
    (Vásquez-Martínez and Phillips, 2000)
    1. Leguminosae
    2. Palmae
    3. Myristicaceae
    4. Euphorbiaceae

III AÑANGU, ECUADOR, floodplain
    (Balslev *et al.*, 1987)
    1. Palmae
    2. Moraceae
    3. Bombaceae
    4. Myristicaceae

IV RIO JURUÁ, AMAZONAS, BRAZIL
    (da Silva *et al.*, 1992)
    1. Leguminosae
    2. Sapotaceae
    3. Lecythidaceae
    4. Chrysobalanaceae

V   RIO CAMANAÚ, AMAZONAS, BRAZIL
    (Milliken *et al.*, 1992)
    1. Leguminosae
    2. Lecythidaceae
    3. Burseraceae
    4. Sapotaceae

VI RONDÔNIA, BRAZIL (Lisboa, 1989)
    1. Leguminosae
    2. Euphorbiaceae
    3. Cochlospermaceae
    4. Moraceae

VII RIO XINGU, PARÁ, BRAZIL (Campbell *et al.*,1986)
    1. Leguminosae
    2. Palmae
    3. Lecythidaceae
    4. Moraceae

VIII CAXIUANÃ, PARÁ, BRAZIL (Almeida *et al.*, 1993)
    1. Leguminosae
    2. Sapotaceae
    3. Moraceae
    4. Lauraceae

IX AMAPÁ, BRAZIL (Mori *et al.*, 1989)
    1. Apocynaceae
    2. Sapotaceae
    3. Mimosaceae (Leguminosae)
    4. Burseraceae

B. TROPICAL ASIA

I   E. GHATS, INDIA (Kadavul and
    Parthasarathy, 1999)
    1. Moraceae
    2. Euphorbiaceae
    3. Verbenaceae

    W. GHATS, INDIA (Ramanujam and
    Kadamban, 2001)
    1. Melastomataceae
    2. Euphorbiaceae
    3. Anacardiaceae

II HALMAHERA–MOLUCCAS, INDONESIA
    (Whitmore *et al.*, 1987)
    1. Myrtaceae
    2. Guttiferae
    3. Lauraceae
    4. Burseraceae

III PAPUA NEW GUINEA (Wright *et al.*, 1997)
    1. Lauraceae
    2. Myristaceae
    3. Moraceae
    4. Meliaceae

TRANSITIONAL FORESTS

These forests are mainly in a belt around the southern fringes of the Amazon rain forest, where the climate is more seasonal and many of the trees are semi-deciduous. The three types of this forest are dominated by different species. To the east it is babaçu palm forest (**251**), while to the west (mainly in the State of Acre, Brazil) the forest is dominated by species of *Bambusa* and *Merostachys* bamboos (**252**). In this forest the bamboos reach high into the other trees for support. The third type of transition is forest dominated by lianas especially of the families Bignoniaceae, Malpighiaceae, and Menispermaceae (**253**). This type of forest is most abundant between the Xingu and Tapajós rivers.

Attalea speciosa Mart. ex Spreng.

This palm, locally known as babaçu (**251**), is characteristic of open forests in the southern fringes of the Amazon forest. In that region the forest is more open, with all trees of about the same height. In addition to babaçu, other species of palm include *Oenocarpus distichus* (bacaba), *Jessenia bataua* (patauá), *Euterpe precatoria* (Açaí da mata), and *Maximiliana regia* (inajá). These species may occur mixed together or one species may dominate. The babaçu often forms large pure stands, especially in the transition forests between the rain forest and the savannas of central Brazil. The area of this palm has increased through deforestation because it is fire resistant. It is also of considerable economic importance to the region because of the oil, charcoal, and other products made from the fruit.

251 *Attalea speciosa* (babaçu palm) is characteristic of transitional forest between the Amazon rain forest and the cerrado or savannas of central Brazil. It occurs in large one-species stands.

252 View of the interior of bamboo-dominated forest, which is hard to penetrate because of the spiny bamboos.

253 Liana forest dominated by woody lianas covering most of the trees.

FLOODPLAIN FORESTS

There are two principal types of seasonally inundated floodplain forests in Amazonia, which are locally termed várzea and igapó. Várzea is on land flooded by white water rivers, where the soil is enriched annually by alluvial matter, while igapó occurs in areas flooded by black or clear water, and generally on a poor sandy soil (Prance, 1979). Both types are less species diverse than in terra firme forest, but várzea is more diverse than igapó. Some floodplain tree species are described below.

Ceiba pentandra (L.) Gaertn.

The silk cotton or kapok tree in the family Bombacaceae is a giant of the white water flooded várzea forests of Amazonia (254–256). It is the tallest tree beside many rivers and has a characteristic umbrella-like crown. The flowers are pollinated by bats and, when mature, its large ovoid fruits are filled by a fibre surrounding the seeds. The fruits split open to release showers of white fluff, which float in the air and disperse the seeds. The young trunks are spinous, but the spines have disappeared by the time the tree matures. The large crotches of the branches of the Ceiba tree are the favourite nesting sites of the harpy eagle. *Pachira aquatica*, another member of the Bombacaceae, is also a common species in seasonally flooded areas. Some other typical species of the várzea forest include *Virola surinamensis*, a much sought after timber tree, *Carapa guianensis* (family Meliaceae), *Hevea brasiliensis* (rubber tree), and several species of *Ficus* (242).

Triplaris surinamensis Cham.

This member of the family Polygonaceae is a typical species of igapó, or black water flooded forest (243). It has hollow branches which are inhabited by aggressive fire ants. Both the tree and the ants are called tachi in Brazil. The attractive large inflorescence starts white, and the flowers gradually turn pink. When in flower it is one of the most conspicuous trees of the igapó. Some other species of this habitat include *Piranhea trifoliata* and *Alchornea castaneifolia* (both in the family Euphorbiaceae), *Copaifera martii* (family Leguminosae), and many species of the Myrtaceae family.

254 Detailed view of the canopy of *Ceiba pentandra*.

255 *Ceiba pentandra* (kapok tree) is typical of the Amazonian floodplain or várzea forests and is easily identified when it emerges above the canopy, by its characteristic umbrella-shaped crown.

256 Massive root buttresses at the base of the trunk of *C. pentandra*.

PERMANENT SWAMP FOREST

Some parts of both black and white water areas are permanently waterlogged. Here, the forest is generally dominated by palms, especially *Mauritia flexuosa* or *Euterpe* species. The soil is mostly a eutrophic humic gley, which is a waterlogged soil lacking oxygen.

Mauritia flexuosa L. f.

The mauritia palm (**257**), called buriti in Brazil and moriche elsewhere, is typical of permanently flooded areas. Where water lies all year around, there are often pure stands of this species. It is a large palm with fan-shaped leaves. The fruits are covered in reddish scales and the pulp underneath them is eaten in many different ways throughout Amazonia. The trees are either male or female. Unfortunately, in some places only male trees remain, since local people have felled the female trees to harvest the fruit in an unsustainable manner. Another much used palm, *Euterpe edulis*, also often grows in swamp forest. This palm is used for palmito or palm heart, and the pulp of the fruit is also a favourite flavour in the Amazon Delta region.

257 *Mauritia flexuosa* (mauritia palm) occurs in swamps throughout Amazonia.

MANGROVE FORESTS

The red mangrove, *Rhizophora mangle* L., is the most typical plant of mangrove forests of South America (**244**). These coastal, salt-water flooded forests occur in coastal areas throughout the tropics. The species of *Rhizophora* are interesting because they produce viviparous seedlings in which the embryo germinates within the fruit while still attached to the tree. When released from the mangrove tree, the seedling floats until stranded on a beach, and anchors itself with roots. The Amazon Delta mangrove forest is dominated by *R. mangle*, which occurs nearest the sea in the saltiest water and is easily recognized by it mass of stilt roots. Further inland, *Rhizophora racemosa* and *R. harrisonii* occur in the less salty water. *Avicennia* species (family Verbenaceae) also occur further inland, and *Laguncularia racemosa* (family Combretaceae) grows in the higher, slightly brackish water.

The Pacific coastal mangrove forests (**245, 246**) are more diverse and, in addition to these genera, include such species as *Mora megistosperma*, *Peliciera rhizophorae*, and *Conostegia polyandra*. Mangrove forests are most important for their role in stabilizing coastlines, and for their rich production of fish and crustaceans. It is tragic that so many have been destroyed all over the tropics. In tropical Asia one of the most important mangrove species is a palm, *Nypa fruticans*. Two other common genera of the Asiatic mangroves are *Bruguiera* (**246**) and *Sonneratia*. There are just over 50 different species of plants around the tropics that have adapted to mangrove forests.

FORESTS ON WHITE SAND

There are various areas in the Guianas and Amazonia where podzols (soils of leached, acidic, white quartz sand) and unconsolidated sand (regosols) occur. The vegetation on this nutrient-poor habitat is distinctive and varies from forest to open savanna. In Guyana this forest is called wallaba forest; this is the local name for the legume *Eperua falcata* Aubl, which is strictly limited to this kind of soil. The inflorescences of this species hang down below the crown on long peduncles, a common adaptation for bat-pollinated species, where the flowers are made easily accessible to the bats. Other common species of wallaba forest include *Catostemma fragrans* and *Licania buxifolia*. There are also considerable areas of white sand in northern Amazonian Brazil, especially in the large area between the Rios Branco and Negro.

In these places the forest is more open and is less species diverse than the terra firme forest, although a number of endemic species occur. For example, *Caryocar gracile* (**258**) is a tree confined to this habitat. The trees are often tortuous and laden with epiphytes (**259**). There is a confusing number of local names for the white sand formation, but the most used ones in Brazil are Amazonian caatinga and campina. Two species of rubber occur in the caatinga; *Hevea rigidifolia* and *H. camporum*. As the name of the first species implies, the caatingas have a xeromorphic aspect with thick leaves, while an abundance of lichens and mosses grow on their branches, and also on the soil surface.

In wetter places on sandy soil in Guyana, another species of legume, *Mora excelsa* Benth., forms large dominant stands. It occupies the low ground beside rivers in the gallery forest floodplain, which is termed mora forest. Other species that occur in this forest include the legumes *Pterocarpus officinalis* and *Pentaclethra macroloba*.

DRY FORESTS

In addition to the humid forest types described above, there are large areas of dry forest scattered throughout the neotropics. The largest area is in the planalto of central Brazil, where savanna (cerrado) and savanna forest (termed cerradão, **249**) are interspersed depending on local conditions. Cerradão is a low, dense evergreen or semi-deciduous savanna forest, in which the trees are 5–15 m tall, and are close but do not form a continuous canopy. Some typical species of cerradão include *Dipteryx alata*, *Magonia pubescens*, *Curatella americana*, *Callisthene fasciculata*, *Pterodon pubescens*, *Roupala montana*, *Tabebuia caraiba*, and *Emmotum nitens*.

Bordering the cerrado, and in some cases the rain forest, much semi-deciduous seasonal dry forest occurs in Brazil and Bolivia. This forest varies in height, but the canopy is often between 15 and 23 m high, with a few scattered emergents. Some of the upperstorey trees include *Apuleia molaris*, *Aspidospermum nitidum*, *Copaifera langsdorfii*, *Enterolobium schomburgkii*, *Hymenea stilbocarpa*, *Vochysia ferruginea*, *Jacaranda copaia*, *Geissospermum sericeum*, *Cenostigma macrophyllum*, *Physocalymma scaberrimum*, *Lafoensia pacari*, *Combretum leprosum*, and *Bowdichia virgilioides*. This type of forest extends in patches into the savanna or cerrado region of central Brazil.

Dry forest occurs in many other places in South and Central America, from Mexico to Argentina; especially in northern Colombia and Venezuela and in some arid Andean valleys.

258 *Caryocar gracile* is a typical species of the northwest Amazonian forest on white sand.

259 Campina forest on white sand near to Manaus, Brazil. In this type of forest the trees are often tortuous and the branches laden with epiphytes.

THE FORESTS OF SOUTHEAST ASIA AND AUSTRALIA

The rain forests of southeast Asia differ from those of Africa and America by the dominance of one tree family, the Dipterocarpaceae (260–262). This large family, with almost 400 species in Malesia, is abundant in the lowland rain forest and swamp forests of the region. Many of the dipterocarps are very tall (*ca.* 60 m) and so these forests are taller than other rain forests. The Malesian rain forest is also interesting because it divides clearly into two, reflecting the Tertiary geological history of the region. The western part is formed by the Sunda shelf (with its characteristic vegetation, 260–262) and the eastern by the Sahel region. Biogeographically, these regions have been divided by Wallace's line (first suggested by the great 19th century naturalist Alfred Russel Wallace), and their exact boundaries have been debated ever since. The Dipterocarpaceae are much more abundant in the Sunda region. All large dipterocarps yield useful timber, which is why so much of the region has been commercially exploited.

The dipterocarps also have the unusual reproductive strategy for tropical trees of mast fruiting, where fruits are produced in vast quantities at intervals of a few years. This strategy is more common in temperate forest species. The dipterocarp seeds germinate immediately, and in a mast year a vast carpet of seedlings grows on the forest floor (see Ashton, 1989 for an account of the reproduction biology).

As in the neotropics, there is a great variety of forest types in the Asian tropics. Areas on white sand are termed heath forests, and a forest unique to the region is the peat swamp forest of Borneo. The mountains of New Guinea and Borneo offer a range of upland montane forests. Some sample trees of the region are described below.

260 Dense canopy formed by a stand of tall trees of *Dryobalanops aromatica* growing in a Forestry Research Station near Kuala Lumpar, Malaysia. This dipterocarp also bears viviparous seedlings. (Photo copyright of Bryan Bowes.)

261 View of the rain forests of Brunei, which are relatively intact compared with the other nations of Borneo.

262 Specimen of the dipterocarp *Shorea agami* growing in a Forestry Research Station near Kuala Lumpar, Malaysia. (Photo copyright of Bryan Bowes.)

Koompassia excelsa (Becc.) **Taub.**

The tualang tree (**263**) is leguminous and the giant of the rain forests of Malesia, even exceeding the dipterocarps in height. It is the tallest known tropical angiosperm, and a tree of 84 m height has been recorded in Sarawak. However, it is not a good timber tree because the wood splits too easily to be of value. The tualang tree occurs in the rain forests of the Malay Peninsula, Sumatra, and Borneo and so is distributed only in the Sunda region. It is also an unusual rain forest tree in that it is deciduous and loses its leaves in the dry season.

Shorea spp.

Shorea is the largest genus of the dipterocarps (**262**), with about 194 species distributed from Sri Lanka and southern India, through the Malay Peninsula, to the Philippines and Molucca Islands. It is the dominant emergent tree genus of the lowland forests of Sundaland, as well as the most important timber genus. The commercial name in Malaya for many species of *Shorea* is meranti. Species of *Shorea* occur in other types of forest such as riverine forest and peat swamps, and a few even occur in montane forest up to 1,750 m (for example, *Shorea brunnescens* Ashton, in the mountains of Borneo). *Shorea macrophylla* and some other species produce little nuts called illipes, which contain a fat used as a substitute for cocoa butter. *Shorea albida* is a characteristic species of heath forests and white sand soil.

Dryobalanops rappa **Becc.**

This is a member of another small genus (**260**) of timber-producing dipterocarps. The species is endemic to Borneo and is characteristic of mixed peat swamp or moor forest. This type of forest, overlying sandy terraces and podzols, is unique to the Malesian tropics in Borneo and Sumatra. *D. rappa* (kapur paya) sometimes forms nearly pure stands, which is another characteristic of some of the peat swamp forests of Borneo. The trunk of *D. rappa* has a shaggy bark, and the seedling leaves are much bigger than the adult ones. Another dipterocarp that dominates some areas of peat swamp forest is *Shorea albida* (alan tree). *Parastemon urophyllus* (family Chrysobalanaceae) is also a typical moor forest species.

Eucalyptus deglupta **Blume**

The genus *Eucalyptus* (**264**) is primarily distributed in Australia (Chapter 4), but *Eucalyptus deglupta* ranges from Mindanao in the Philippines, through Sulawesi, the Moluccas, and New Guinea, to

263 *Koompassia excelsa*, one of the largest trees of the forests of Borneo.

264 Specimen of *Eucalyptus urnigera* growing in Australia. (Photo copyright of Bryan Bowes.)

New Britain and New Ireland. It therefore represents the species of the Sahal region east of Wallace's line. This is the only true rain forest *Eucalyptus*, and other tropical species occur in savanna woodlands of Malesia. *E. deglupta* is a colonizing species on eroded slopes and abandoned fields. In New Guinea it is known as kamerere. Trees of 78 m in height and 7 m girth have been recorded in the Philippines. Since *E. deglupta* is fast growing and has a good timber, it has been widely planted in other parts of the tropics, both as a commercial timber and as an ornamental tree.

Agathis

The genus *Agathis* is interesting because it is the most tropical of all conifers with many species being true rain forest ones. The genus is distributed from Sumatra and Malaya, through Indonesia, to Queensland and across the Pacific to Fiji and New Caledonia. A single species (*A. australis*) also occurs in the North Island of New Zealand. The species of *Agathis* are characteristic of forests without a marked dry season. Two species of *Agathis* occur in the white sand heath forest over podzols, namely *A. borneensis* in Borneo and *A. labillardierei* in New Guinea. They often occur in large pure stands in the heath forest, a trait which is quite common in other heath forest species. *Agathis* furnishes a high-grade timber and is also much exploited for copal resin. Some trees of *A. microstachya* in the forests of Queensland are estimated to be 1,000 years old.

Tectona grandis L. f.

This species is the source of teak wood. It grows in the monsoon forests of Myanmar, India, Thailand, and Indonesia. These forests suffer a long dry season and, like many other monsoon forest trees, teak loses its leaves for many weeks in the dry season. Teak cannot be cultivated in ever-wet climates but is widely cultivated in areas where the rainfall is more seasonal. *T. grandis* is a large and fast growing tree that can reach 45 m in height. Its straight-grained wood feels oily to touch and is a favourite for garden furniture, window and door frames, and in boat building.

THE FORESTS OF AFRICA

African forests occur in an equatorial belt across the continent. There are two gaps in regions with drier climate in the Togo–Benin region of West Africa (formerly known as the Dahomey Gap) and in East Africa (**265**). The forests of Africa have a similar physiognomy to those of South America but, because of the long isolation, very few species are common to both continents. *Parinari excelsa* (family Chrysobalanaceae) and *Symphonia globulifera* (family Clusiaceae) are two forest species which occur on both continents. The dipterocarps are common in the forests of southeast Asia, rare in the savanna forest of Africa, and are only represented by a single species in the Guayana Highland of South America. The African forests are also poorer in the number of tree species in comparison to Asia and America. This is probably partially due to the extent to which the species composition in Africa was reduced during periods of drier climate in the Pleistocene. The richest forests occur today in Cameroon, Gabon, and the Eastern Congo Republic – these are areas that remained as refugia during the periods of dry climate. Dominance of the forest canopy by a single tree species is a common feature of the lowland evergreen forests of Africa, especially by *Brachystegia laurentii*, *Gilbertiodendron dewevrei*, and *Cynometra alexandri*. The northeastern part of Madagascar was also covered with rain forest and today is of great importance for

265 Miombo woodland in Malawi, Africa, dominated by species of *Brachystegia* and *Parinari curatellifolia*. This semi-deciduous woodland covers much of eastern Africa.

the high amount of endemism in the species still remaining there (**266–268**). The conservation of the remnants of this unique forest is of the utmost importance. Some trees characteristic of African forests are described below.

Gilbertiodendron dewevrei (De Wild.) J. Leonard

This member of the family Caesalpiniaceae is one of the commonest trees of African lowland forests, and ranges from the forests of Nigeria to Angola. It exhibits an extraordinary dominance in the forest canopy over large areas of the northern and eastern rim of the Congo River basin. It sometimes comprises 70% or more of the large trees over

extended areas. The stands of *Gilbertiodendron* forest are interspersed with a matrix of mixed forest where it is absent. No differences in topography or soil type have been found between forests with and without *G. dewevrei*, so the distribution of this species remains a mystery. The canopy formed by *G. dewevrei* is dense, and consequently little ground vegetation occurs underneath it. The tree is usually 30–40 m tall, and can reach 1.5 m in diameter. Half to two-thirds of the height is occupied by the tree's wide crown, which branches from low down. In *Gilbertiodendron* forest, a few other taller species, such as *Tieghemella heckelii* (family Sapotaceae) and *Oxystigma oxyphyllum* (family Caesalpiniaceae), can emerge above the canopy level.

Cynometra alexandri C. H. Wight

The lowland forests of the Congo region may also be dominated by *Cynometra alexandri*, which is another member of the Caesalpiniaceae family. Known as ironwood or mahimbi, this species has a useful timber. It is a canopy species and in a forest inventory at Edoro in the Congo Republic, mahimbi comprised 39% of the basal area of stems. Ironwood is especially abundant at altitudes of around 1,000 m in northeastern Congo and into Uganda. It is thus more characteristic of lower montane forest, whereas *Gilbertiodendron* is more

266 View of the Madagascan rain forest. Most of its species are endemic to the island, yet the forest is disappearing at an alarming rate.

268 Madagascan rain forest, showing a rare endangered species of palm.

267 Another view of the Madagascan rain forest.

common in lowland forest, especially near to rivers. Specimens of *Julbernardia seretii* (another caesalp) and *Staudtia stipitata* (family Myristicaceae) are often present in the forest as co-dominants with *C. alexandri*.

Carapa procera DC. and *C. grandiflora* Sprague

These species are members of the mahogany family and are known as African crabwood. They occur from Sierra Leone to Uganda and are widely distributed as far south as Angola. The ranges of these two species of *Carapa* overlap in the Congo and Zaire. Their wood is much used for furniture, flooring, and in Uganda for mine timber. *C. procera* is a typical tree species of swamp forest and stream valleys in western Nigeria, but it also occurs on well-drained upland sites in Cameroon and elsewhere. It is very closely related to the South American species *C. guianensis* from which it is doubtfully distinct.

Podocarpus falcatus R. Br. ex Mirb.

This is one of the few conifers of the forests of tropical Africa. It occurs in areas of drier montane rain forest in Ethiopia, through East Africa, to the Cape. It grows up to 30 m tall with characteristic bluish foliage that has made it popular in cultivation. *P. latifolius* occurs in the wetter montane forests of Cameroon, southeast Nigeria, and over to East Africa. These two species of *Podocarpus* are sometimes placed in a separate genus *Afrocarpus* (see also Chapter 4). Other tree species of the African montane forest include *Entandophragma excelsum*, *Ficalhoa laurifolia*, *Ocotea usambarensis*, and *Strombosia grandifolia*. *Symphonia globulifera* (family Clusiaceae) grows in both the montane and the lowland forests of Africa and is one of the few species to occur on more than one continent, since it is also found in the forests of South America.

Brachystegia species

This is another genus of the Caesalpiniaceae family, with about 30 species. *B. laurentii* often exhibits single species dominance in the rain forests of the Congo Basin and grows to a height of 45–50 m, while its trunk often exceeds 1.5 m in diameter. Species of *Brachystegia* are also major components of the deciduous miombo woodlands of South Central Africa, where collectively they dominate this

type of vegetation. For example, *Brachystegia taxifolia* forms almost pure stands towards the upper limits of the miombo forests on the Nyika Plateau of Malawi. Some species of *Brachystegia* are used for their timber under the name okwen.

Diospyros species

The ebony genus *Diospyros* has many species in the forests and savannas of Africa and South America. *D. crassiflora* is a lowland rain forest species of Africa, and has the darkest of all heartwoods, that contrasts with the pale sapwood. Black ebony wood is a favourite for carving and for many specialized uses, such as knife handles and butts for billiard cues. *D. mespiliformis* occurs in forests fringing rivers, while *D. chevaleiri* is common in the evergreen rain forests of Ghana. *D. xanthochlamys* is one of the most abundant trees of the South Bakundu Forest Reserve in Cameroon. Species of *Diospyros* are also frequent in the lowland mixed dipterocarp forests of Malesia.

OLD WORLD DRY SEMI-DECIDUOUS FORESTS

These forests border many of the rain-forest areas where the climate is drier and more seasonal. These semi-deciduous forests often form the transition region between rain forest and savanna or dry deciduous forest types. They may also occur on certain soil types. Extensive areas of this formation occur in Africa and some in Malesia, as well as in some areas of Thailand, Vietnam, Myanmar, and eastern Australia.

In Malesia there is a decrease in the number of dipterocarps in such forests, but *Anisoptera oblonga* is a characteristic dipterocarp. Other common species are *Tetrameles nudiflora* (family Datiscaceae) and *Garuga floribunda* (family Burseraceae). In Africa some of the common savanna trees enter the semi-deciduous forest belt; for example, *Borassus aethiopum* (a palm) and *Afzelia africana*. The latter is a member of an important timber-producing genus with a distinctive wood, pale straw-coloured sapwood, and rich red-brown heartwood, which is much sought after for joinery.

MONTANE FORESTS

Various types of these forests occur in all parts of the tropics where mountains and even quite low hills occur. Montane forests are usually divided into lower and upper montane forests, while the more humid cloud forest, elfin, or sub-alpine forests occur at higher levels. Many small patches of montane forest occur on lower hills due to the Massenerhebung effect (**269**; see also Grubb and Whitmore, 1966). The altitude at which the different montane types occur depends on both latitude and local climatic conditions. In the Cajamarca region of Peru and nearby Colombia, a forest dominated by *Podocarpus oleifolius* occurs. In Uganda there is also a *Juniperus/Podocarpus*-dominated dry montane forest. A good summary of African montane forests was given by Hamilton (1989). In the Malesian mountain flora, many plant families of largely temperate region distribution occur, such as the Aceraceae, Fagaceae (**270**), and Podocarpaceae.

269 Submontane forest on the slopes of Etinde Mountain in Cameroon.

SECONDARY FORESTS

Unfortunately, secondary forest is on the increase where the original forest has been felled and the area then abandoned. This has happened on all three major rain forest continents. Secondary forests consist of pioneering species that naturally occur in gaps in primary forest. There are three look-alike genera, one for each continent, that often dominate these areas, and all have large, often palmate, leaves. *Cecropia* species (**271, 272**) dominate this habitat in South America, and have associated ants that inhabit the hollow trunk and branches. *Musanga cecropioides* (**273**), also in the family Cecropiaceae, often dominates African secondary forests, while *Macaranga* (family Euphorbiaceae) is the equivalent in Malesia. Secondary forest species are light demanding and intolerant of shade. They have efficient methods of seed dispersal; for example by bats in the case of *Cecropia*. They are fast growing and often exclude other species, but since they are short-lived, space gradually opens up for the more shade-tolerant primary forest species. *Musanga cecropioides* (**273**) grows to 11 m in three years and to 24 m in only nine years. Another fast-growing pioneer species of tropical America is *Ochroma lagopus* (balsa), which can reach a height of 18 m in five years. It forms a soft, light wood that has been much used for rafts and model airplanes. Another important pioneer species of Malesia is *Adinandra dumosa* (family Theaceae), which begins to flower at a height of 2 m when it is only two years old and

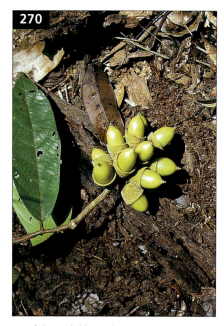

270 Acorns of the oak-like *Lithocarpus*, which is a genus abundant in the forests of Malesia. The Fagaceae is an example of one of the otherwise predominantly temperate forest families found in the region.

continues to produce flowers continuously throughout its life.

Species of *Trema* (family Ulmaceae) are frequent in secondary forests of America and Asia. A common large tree in Africa is *Milicia excelsa* (Moraceae), the wood of which is a popular substitute for teak, and is much used for wood carving and parquet flooring.

271, 272 Species of *Cecropia* in South American secondary forest. When the original rain forest is cut *Cecropia* invades rapidly, as the seeds are dispersed by bats and birds, and often forms pure stands. The hollow trunks and branches of *Cecropia* are occupied by aggressive fire ants.

273 *Musanga cecropioides*, a characteristic species of secondary forest in Africa. It resembles the genus *Cecropia*, which occupies the same habitat, but unlike *Cecropia* it does not have ants in the trunk.

SECTION 3 TREE MORPHOLOGY, ANATOMY, AND HISTOLOGY

CHAPTER 6

Woody thickening in trees and shrubs

Bryan G Bowes

INTRODUCTION

BROADLEAVED TREES AND CONIFERS

In conifers and broadleaved trees undergoing woody thickening (**274**), new (secondary) conducting vascular tissues are generated from a specialized hollow cylinder of tissue, termed the vascular cambium (**275**). Secondary wood (xylem) is formed to the inside of the cambium, while phloem (inner bast) is developed on the outside (**276D, 277D**). Nearly all of the increasing girth of such a tree is due to its expanding core of wood, which provides both water conduction and mechanical support.

Only a relatively thin investment of secondary phloem is formed (**275**). Its primary function is to transport sugars and other nutrients in solution, but in some tree species, limited mechanical support is also provided by phloem fibres. Accompanying and accommodating this internal thickening of the vascular tissues, the original epidermis is replaced by a layer of bark (**275, 278**) which encloses the trunk, branches, and older roots of the tree. In some trees, such as the Californian redwoods (*Sequoia sempervirens* and *Sequoiadendron giganteum*) and

274 Massive ancient pollard of the broadleaved deciduous *Fagus sylvatica* (beech) growing in England.

275 Cross-cut trunk of *Quercus petraea* (sessile oak), showing numerous rays and wide secondary xylem with dark red heartwood (1). Positions of vascular cambium (2), phellogen (3), secondary phloem (4), and bark (5).

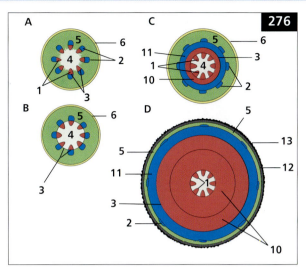

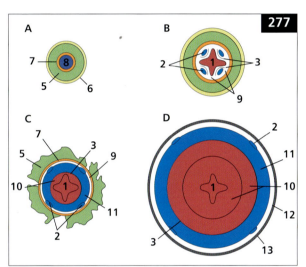

276, 277 Diagrammatic representation in TS of stages in the secondary thickening of a broadleaved species woody stem (**276A–D**) and root (**277A–D**). Primary xylem and phloem (1, 2), vascular cambium (3), pith (4), cortex (5), epidermis (6), endodermis (7), procambium (8), pericycle (9), secondary xylem and phloem (10, 11), cork cambium (12), and cork/bark (13).

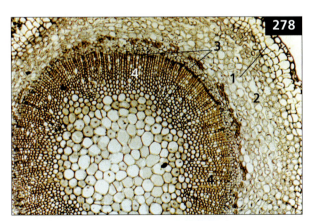

278 TS of a thickened twig of *Ligustrum vulgare*. Cork (1), cortex (2), position of vascular cambium (3), and secondary xylem (4).

279 Cross-cut trunk of a stringy-bark *Eucalyptus* sp. Secondary phloem (1) and stringy bark (2).

280 A large specimen of *Dracaena draco* (dragon tree), an arborescent monocotyledon indigenous to Tenerife and the Canary Islands.

species of *Eucalyptus* in Australia, the bark can become very thick and fire-resistant (**279**). Bark initially consists of a cork layer developed from a cork cambium (phellogen, **276D, 277D, 278**); but at a later stage, tissues may also be incorporated from the older and non-functional secondary phloem (see below).

ARBORESCENT MONOCOTS

A number of monocots undergo an unusual form of woody thickening, and some may form quite large trees. This is the case with various liliaceous species, such as *Dracaena draco* (dragon tree, **280**, which may grow to 15–20 m tall), *Yucca brevifolia* (Joshua tree), *Y. elephantipes*, and *Aloe dichotoma* (quiver tree). Members of the agave family, such as *Cordyline australis* (giant dracaena, which grows up to 10 m tall) and *Nolina curvata* (**281**), also form large trees. Various species of *Pandanus* (screw pine) and *Xanthorrhoea* (grass tree) are arborescent, but generally do not grow very tall.

In the older stems of these monocot trees, a cylindrical cambium (thickening meristem) develops externally in the parenchyma of the outer cortex. This then cuts off new individual secondary vascular bundles to its inside (**282, 283**), with each bundle consisting of xylem surrounding a strand of phloem.

281 *Nolina curvata*, a monocotyledonous tree belonging to the family Agavaceae.

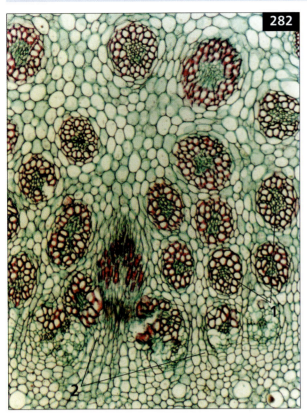

282 TS of the young stem of the monocot *Dracaena* sp. showing the formation of secondary vascular bundles (1) from the thickening meristem (2).

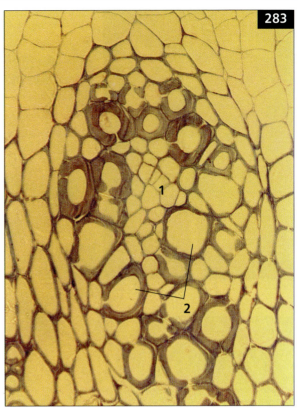

283 TS of *Dracaena* sp. stem showing detail of a secondary vascular bundle. Note the central strand of phloem (1) surrounded by large and thick-walled tracheary elements (2).

(This mode of secondary thickening contrasts greatly with that in conifers and broadleaved trees, as outlined above.) As the monocot trunk thickens, a cork cambium forms a layer of bark to replace the original epidermis.

VASCULAR ACTIVITY IN BROADLEAVED AND CONIFEROUS TREES
CONDUCTING TISSUES OF THE YOUNG STEM AND ROOT

In the young leaves of an actively growing bud, the precursor of the vascular tissues is composed of individual, longitudinally orientated, procambial strands. These strands link, at the base of the bud, with the older conducting tissues of the primary (first-formed) xylem and phloem. The elongated procambial cells are densely staining and contrast markedly with the adjacent, lighter staining, future pith and cortical cells (284). In the young elongating

284 RLS of a bud of *Pinus* sp. Shoot apex (1), young leaves (2), and procambium (3).

285 Bursting bud of *Aesculus hippocastanum* (horse chestnut). Note the hairy foliage leaves unfurling and the dark brown burst-open bud scales at the base of the new shoot.

286 Hand-cut TS of the very young shoot of *Aesculus hippocastanum* (horse chestnut) newly emerged from its bud (*cf.* **285**). A closed ring of primary vascular bundles (1) surrounds the central pith (2). Cortex (3), young fibres (4), and epidermis (5).

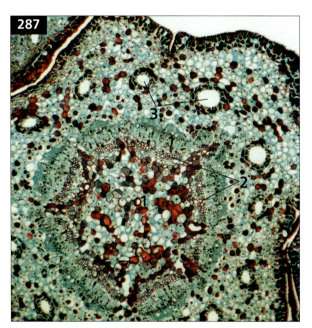

287 TS of an unthickened stem of *Pinus* sp. showing pith (1) surrounded by a ring of primary vascular bundles (2). Note resin canals (3) prominent in cortex.

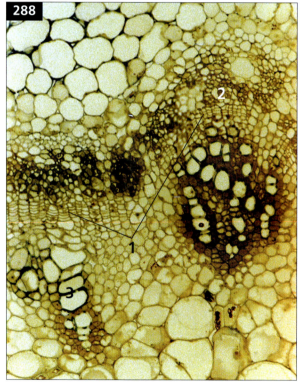

288 TS of an old *Phaseolus vulgaris* stem showing the fascicular vascular cambium (1), primary phloem (2), and xylem (3).

twig which develops from the bud (**285**), a ring of individual veins or vascular bundles differentiates from the procambial strands (**276A, 286, 287**). Each vascular bundle consists internally of a thick deposition of primary xylem, which lies adjacent to the pith, while a thinner layer of phloem forms externally next to the cortex. These vascular tissues remain separated by a layer of fascicular cambium, which represents a persistent residue of the procambium (**288**).

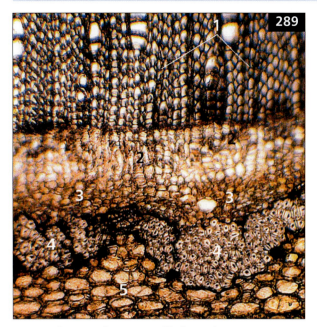

289 Hand-cut TS of a young, still-elongating, first-year stem of *Aesculus hippocastanum* (*cf*. 286) showing the early onset of secondary thickening in an arborescent species. Secondary xylem (1), secondary phloem (2), primary phloem (3), fibres (4), and cortex (5).

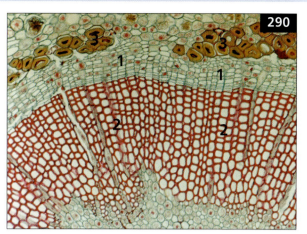

290 TS of a woody twig of *Ginkgo biloba* (maidenhair tree) with a well-developed vascular cambium (1), secondary xylem (2), and phloem fibres (3).

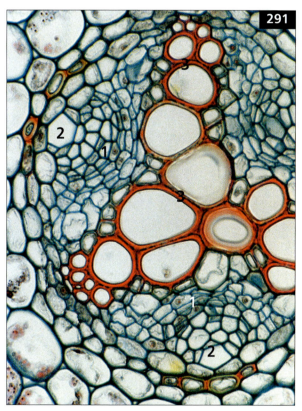

291 TS of a *Ranunculus* sp. root showing vascular cambium (1) lying between primary phloem (2) and xylem (3).

Secondary thickening in a twig normally commences before a young internode (the length of stem between successive leaves) has fully elongated (289). The fascicular cambial cells become active and divide in a plane tangential to the surface of the stem (288). The parenchyma cells situated between the individual vascular bundles also become involved, and a continuous ring of vascular cambium is formed (276B, 286, 287). This generates a ring (as seen in transverse section, 276C) of secondary xylem internally and phloem externally (290).

In a young tree root, a central cylinder of procambial tissue is formed at its tip (277A). Later, in the outer procambium, alternating longitudinal strands of primary phloem and narrow xylem elements are differentiated, while the central tissue frequently differentiates into a core of wider-diameter, primary xylem elements (291). The earlier-formed peripheral xylem elements show as arms radiating from this core, while the strands of phloem are located between the ridges (277B). In the thickening root, the vascular cambium initially arises in the residue of procambial tissue lying between the phloem strands and the fluted xylem core. The procambial cells divide tangentially, and cut off secondary xylem internally and phloem externally (277B–C). Subsequently, this cambial activity

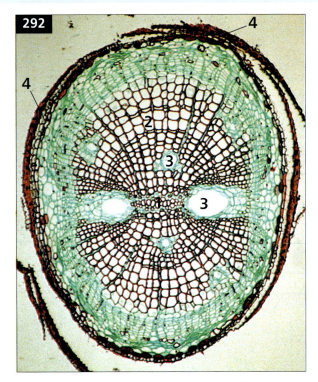

293 Cross-cut trunk of *Larix kaempferi* (Japanese larch) showing its annual rings.

292 TS of a *Pinus* sp. root showing early secondary thickening. Primary and secondary xylem (1, 2), resin canal (3), and bark (4).

294 Coring the trunk of a species of *Betula* (birch). A radial wood cylinder is extracted by this apparatus, and its surface shaved flat to reveal its annual rings under a magnifying glass.

spreads sideways to cover the primary xylem arms, so that the cambial layer initially has a convoluted outline in a cross section of the root. It soon becomes circular (**277C**), due to the uneven development of secondary xylem filling the original primary xylem flutes. However, the large resin ducts present in some conifer roots may distort the development of this xylem (**292**).

CAMBIAL ACTIVITY AND PERIODICITY

In temperate and boreal trees, the vascular cambium becomes dormant in winter. This is reflected in the growth rings which are usually evident in a cross-cut tree trunk. These growth increments typically form annually (**275, 293**), and generally allow an estimate of the age of the trunk. However, this measure of age is only valid in a trunk up to about 1.5 m above the ground; the level where the girth of an intact tree trunk is normally measured. This position is assumed to be the height of a tree sapling after one year of growth. A coring device can also be used to extract a radial cylinder of wood from the living tree to determine its age (**294**). It sometimes happens that

more than one growth ring forms in a year when growth has been discontinuous due to factors such as frost damage, flooding, drought, or defoliation from caterpillar attack. Also, growth rings are often absent in tropical trees (Longman and Jenik, 1987); but nevertheless apparently occur in many indigenous trees of the Amazon Basin and India (Thomas, 2000).

In various broadleaved trees, such as species of *Quercus* (oak), *Ulmus* (elm), *Castanea* (sweet chestnut), and *Fraxinus* (ash), growth rings are very obvious to the naked eye in a cross-cut trunk (**275**). The wood of such trees is termed ring-porous.

295 TS of the ring-porous wood of *Fraxinus americana* (white ash). Note the very wide vessels (1) in the early wood, and the few narrow, single/aggregated vessels (2) in the late wood. Thick-walled fibres (3) and fibre-tracheids (4).

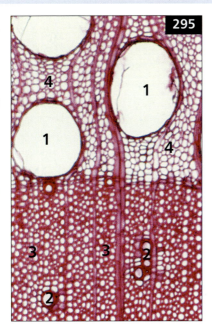

296 TS of the diffuse-porous wood of *Liriodendron tulipifera* (tulip tree) showing the larger vessels distributed fairly evenly throughout a growth ring.

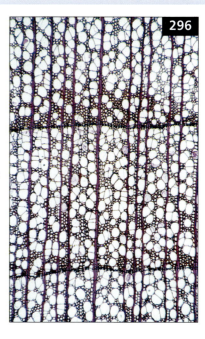

The appearance is due to the formation of abundant wide vessels (dead water-conducting tubes, from several to many cells long) in the newly formed spring wood. These are sharply demarcated from the narrower and thicker-walled conducting elements of the later season's wood (**295**). However, in the more numerous diffuse-porous broadleaved trees, such as species of *Aesculus* (horse chestnut), *Magnolia*, and *Liriodendron* (tulip tree), the larger vessels are more evenly distributed throughout a single year's growth increment. Consequently, growth rings may not be so clearly distinguishable to the naked eye, although they are very evident under the microscope (**296**). In conifer wood, only narrower tracheids (dead single cells) are present. Nevertheless, various conifers develop extensive thicker-walled tracheids in the late wood (**297**), so that rings are often clearly visible in cut trunks (**293**).

297 TS of the wood of *Thuja plicata* (western red cedar) showing its radial arrangement of tracheids. The wide lumina of the early tracheids contrast with those of the narrower, thicker-walled elements in the late wood.

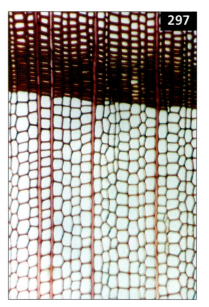

At bud burst or flushing of temperate trees in spring (**285**), the previously dormant vascular cambium becomes activated by auxins, which are hormonal substances synthesized in the newly expanding leaves. The hormonal signal moves from the buds downwards, via the vascular cambium, first into the twigs, then the branches and trunk, until finally reaching the roots. In ring-porous trees, the transmission from canopy to roots takes only a few days, but in diffuse-porous trees it may occur over several weeks, and in conifers the rate of movement is generally intermediate. In a tree suffering environmental stress (Mattheck and Breloer, 1994), cambial activation may not reach the base of the trunk, or is absent on one side of the tree. Consequently, that year's growth increment may be absent or unevenly developed in this part of the tree.

The cambium is composed of cells of two types, termed the ray and fusiform initials. The elongate fusiform initials either lie in horizontal layers (when viewed in a tangential longitudinal section, **298A**) and form a storied cambium, or are more randomly arranged (**298B**). The fusiform cells divide in a predominantly tangential longitudinal plane (**299**), to produce the radial rows of secondary xylem and phloem seen in cross sections of stem or root (**289, 290, 292, 295–297**). However, the fusiform initials also divide in a radial longitudinal plane, to accommodate the increasing circumference of the tree. The ray initials give rise to the radially running rays in which water and nutrients are transported across the axis of the tree (**275, 289, 290, 295–297, 300**). With the increasing girth of the tree, additional rays may be formed following the transformation of fusiform into ray initials.

In broadleaved trees the fusiform initials give rise to vessels and tracheids (collectively termed

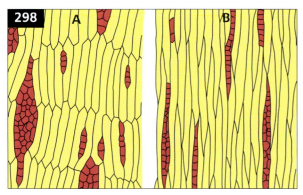

298 Diagrammatic TLS of a storied (**A**) and non-storied (**B**) vascular cambium. (Red indicates ray initials and yellow fusiform initials.)

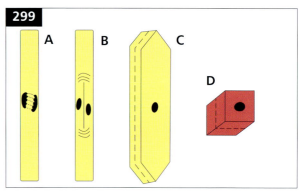

299 Elements of the vascular cambium as seen in diagrammatic longitudinal views. A fusiform initial (yellow) is shown in radial view undergoing mitosis (**A**) followed by cell division (**B**). The daughter cells are seen in oblique tangential aspect (**C**). The ray initial (**D**, in red) has just completed a tangential division.

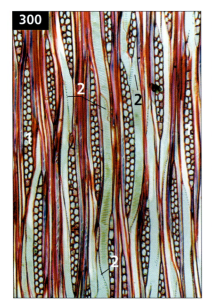

300 TLS of the xylem of *Magnolia grandiflora* (evergreen magnolia). Note the numerous multiseriate rays (1) and scalariform perforation plates (2) in the uniform-diameter vessels.

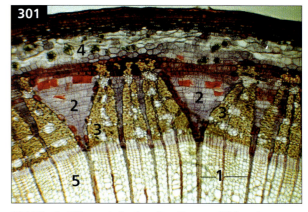

301 TS of a *Tilia* sp. twig showing the rays (1) forming expansion tissue (2) in the secondary phloem. Note the thick-walled fibres (3) in the phloem. Cortex (4) and secondary xylem (5).

tracheary elements) in the secondary xylem, which conduct water from root to shoot. The fusiform initials also give rise to thick-walled fibres (which are dead but lend additional mechanical strength to the tree), as well as living, thin-walled, general-purpose parenchyma cells. Finally, the fusiform initials form the secondary phloem, which is composed of the living, nutrient-transporting sieve tubes (with their associated companion cells), fibres, and parenchyma cells (**301, 302**). In gymnosperms the conducting elements are simpler, with only xylem tracheids and phloem sieve cells occurring (**290, 292, 297, 303**), but fibres and parenchyma cells are present in both tissues (**290**). In conifers resin ducts occur commonly throughout the xylem and phloem (**287, 292, 304**) and resin is often secreted profusely from wounded tissue (**305**).

302 TEM of a leaf vein of *Sorbus aucuparia*: fibre (1), tracheary element (2), parenchyma cell (3), sieve tube (4).

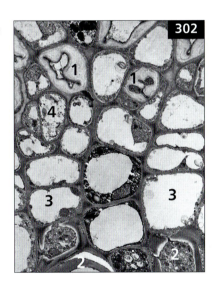

303 RLS of xylem tracheids in *Araucaria angustifolia*. Note the interdigitating pointed tips and numerous bordered pits arranged in two or more vertical rows on the tracheid walls.

304 TS of a twig of *Pinus* sp. showing three annual rings in its secondary xylem. Note the resin canals (1) present in both the xylem (2) and cortex (3).

305 Cut surface of a felled trunk of *Araucaria araucana* (monkey puzzle) covered with congealed resin which has exuded from its numerous severed resin canals.

STRUCTURE AND FUNCTION OF SECONDARY XYLEM (WOOD)

In the wood of a living tree the tracheary elements are dead, and their thick secondary walls are strengthened and waterproofed by impregnation with a complex carbohydrate polymer, termed lignin (Chapter 7). These elements are only permeable to water at their pits, where the secondary wall is absent and the unlignified primary wall is exposed (306). Such pits are generally bordered (307), with the primary wall forming a thin pit membrane, which separates adjacent tracheary elements. Half-bordered pits occur where tracheary elements link to parenchyma cells (306).

Tracheids are single, elongated, and pointed cells with thickened walls, which range from about 15 to 80 μm in width (303). They are characteristic of conifers but also occur in the wood of broadleaved trees. In *Araucaria cunninghamii* (hoop pine) a tracheid may be up to 11 mm in length, but in other conifers they are shorter and often only 1 mm long. The angular tips of tracheids overlap each other (303) and link with adjacent cells by a single to several vertical rows of bordered pits (308). In many conifers the central part of the pit membrane is thickened to form a biconvex torus (309). This probably acts as a valve to block the pit if an embolism occurs, and helps prevent its spread through

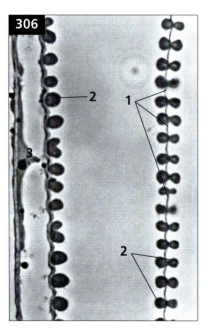

306 LS of a primary xylem vessel showing its thin, unlignified primary wall (1) and the multiple bordered pits formed by thickening of the lignified secondary wall (2). Note also the living parenchyma cell (3), without pits on its side of the compound primary cell wall.

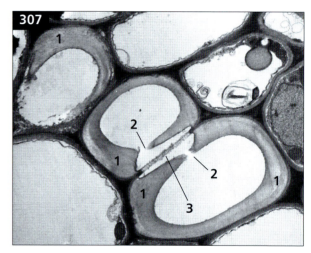

307 TEM of adjacent vessels in the mid-rib of a leaf of *Sorbus aucuparia* (mountain ash). Note the thickened secondary walls (1) and the bordered pit (2) where the unthickened primary wall (3) is exposed.

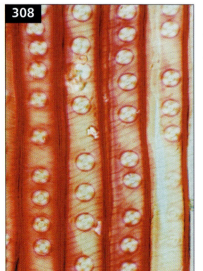

308 RLS of xylem tracheids in *Pseudotsuga taxifolia*, showing bordered pits in the walls.

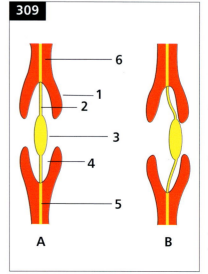

309 Diagram of a conifer bordered pit in a non-cavitated tracheid (**A**), and after cavitation (**B**). Pit border (1), pit membrane/primary wall (2), torus (3), pit cavity (4), middle lamella/primary wall (5), and secondary wall (6).

the adjacent, still-functional tracheids (see below).

Both vessels and tracheids are present in the wood of the vast majority of broadleaved species, but in *Drimys winteri*, *Tetracentron sinense*, and *Trochodendron aralioides*, only tracheids occur. Vessels are formed from two to many elongated cells, with flattened and perforated end walls, joined together in a long file. These cells are generated by the fusiform initials of the vascular cambium (**310A**) and, during the subsequent differentiation of the vessel elements, their cytoplasmic contents degenerate (**310B**). Their abutting end walls partly or completely disintegrate (**310B–C**), while their side walls bear numerous bordered pits (**307, 310C**). In species of *Quercus* (oak) and *Fraxinus* (ash) individual vessels may extend for a metre or more along the axis of the tree, but in *Acer saccharum* (sugar maple) they do not exceed about 30 cm in length (Mauseth, 2003). Vessels are often wider than tracheids (**295**), with some vessels up to 360 μm in diameter. Water moves from tracheid to tracheid via their common pits, where only the unlignified and permeable primary wall is present.

Vessels, although generally longer than tracheids, are not of infinite length, and water must pass from vessel to vessel via the pits in their side walls.

Water movement through a tree (transpiration) occurs though the active xylem (sapwood) and passes from root to shoot. Movement is much more rapid along a wide vessel than a narrower tracheid. It is powered by evaporation of water vapour, via the stomata, from the leaf surfaces, and the consequent pull exerted on the water columns in the individual tracheary elements. In some species of *Eucalyptus* and temperate conifers such as *Sequoia sempervirens* (coastal redwood), water columns are pulled up by leaves in a canopy which may be up to some 100 m above ground level. In certain deciduous trees, such as species of *Acer* (maple), *Betula* (birch), and *Juglans* (walnut), a positive pressure develops in the wood some weeks prior to the buds bursting. This root pressure is due to the secretion of sugars and minerals into the xylem elements. Water then enters the xylem by osmosis and the sugary sap is forced up the tree

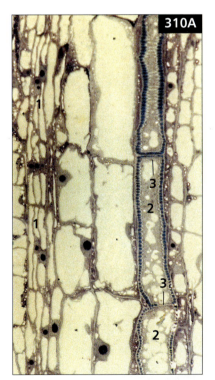 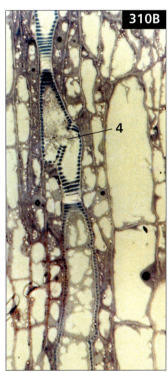 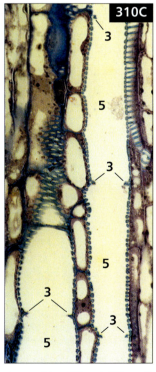

310A–C LMs of the old and woody stem of *Phaseolus multiflorus* (runner bean): **A**, vascular cambium (1) and differentiating vessel elements with thickened, pitted secondary walls (stained bright blue), but with cytoplasm (2), and end walls (3) still apparent; **B**, end wall between two vessel elements has perforated (4); **C**, the elements in two wide-diameter vessels now have fully perforated end walls with only their rims (3) remaining, while the lumens (5) are without cytoplasm.

(Thomas, 2000). In the northeastern USA, the sap of *Acer saccharum* (sugar maple) is tapped and boiled down to produce maple syrup.

The wide vessels characteristic of many broadleaved trees bring attendant dangers. The cohesion of the water molecules in the transpiration columns is reinforced by the adhesion of the molecules to the walls of the tracheary elements, but this effect is negligible at the centre of a wide vessel. Consequently, buffeting by a gale is much more likely to cause rupture of water columns (cavitation) in vessels of the branches and twigs of an oak tree, than in the much narrower tracheids of a pine tree. Cavitation is an embolism, consisting of water vapour under reduced pressure, and it blocks the entire length of an affected vessel or tracheid to water movement (**311**). However, passage of the embolism to adjacent tracheary elements is blocked by their common pit membranes.

In some temperate trees the sapwood becomes cavitated and ceases to conduct water by the end of its first growing season. In others, such as *Robinia psuedoacacia* (false acacia, black locust), the sapwood functions for two to three years, while in *Juglans nigra* (black walnut) it may be active for up to 20 years (Mauseth, 2003). In tropical broadleaved trees, the tracheary elements may remain uncavitated and conduct water for many years (Thomas, 2000).

In palms, no secondary xylem is formed and yet the primary xylem continues to function for water transport for decades, and sometimes even centuries, during the life of the tree (Tomlinson, 2003).

Initially, the main trunk of a tree usually develops a more or less upright habit, but its branches typically grow more or less horizontally or obliquely outwards from the trunk (**312**), and these branches form compression or tension (reaction) wood, each with its distinctive anatomy. In a wind-blown tree, the subsequent straightening of its trunk or branches into an upright position (**313**) involves the formation of reaction wood. Compression wood is formed on the lower side of a conifer branch (**314**) or leaning trunk. This involves an increased production of xylem, which shows as wider growth rings, containing a high proportion of shorter, thick-walled, and heavily lignified tracheids (Thomas, 2000).

By contrast in many broadleaved trees, tension wood is said to characterize the upper side of a leaning trunk or branch. Nevertheless, in some species the growth rings clearly show greater development on the lower side of the branch or trunk (**315**, cf. **314**). In tension wood, fewer and narrower vessels occur than in normal xylem, but numerous unlignified and cellulose-rich gelatinous fibres are present. In felled trees, reaction wood is best avoided for use as timber. Tension wood splits

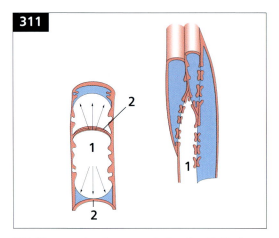

311 Diagram showing an embolism (cavitation) blocking water flow (blue) through a vessel (left) and tracheid (right), the pitted walls of which are shown in red. Arrows show the spread of the embolism (1) through the entire vessel. Perforated rim between vessel elements (2).

312 A specimen of the deciduous broadleaved *Tilia* x *europaea* (lime) in its winter aspect in Scotland. Note the bushy lower trunk caused by the sprouting of numerous epicormic buds.

313 Prostrate wind-thrown large trunk of *Populus* sp. (poplar) in Scotland, which still retains some active roots on its upturned root plate, with three branches having grown up into mini-trees.

314 Large side branch scar on the trunk of *Cedrus deodara* (deodar) with the reaction (compression) wood forming the thicker lower side of the eccentric xylem. Original position of pith (1) and wound callus (2). Note also the coarse and ridged bark on the main trunk.

315 Cut-off trunk from an overgrown coppiced broadleaved specimen of *Castanea sativa* (sweet chestnut). Note how the annual rings are eccentric and much wider on the lower side of the trunk (despite being a dicot), as in the conifer *Cedrus deodara* in 314. Position of original pith (1).

316 TS of *Robinia pseudoacacia* (false acacia) ring-porous wood showing numerous tyloses (1) filling the lumens of the large non-functional vessels. Note also the clustered, thick-walled fibres (2) and rays traversing the wood (3).

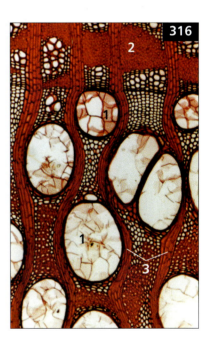

easily on drying, whereas compression wood is hard but brittle to work. On wind-swept specimens of *Pinus sylvestris* (Scots pine) 20–50% of their bulk is represented by compression wood (Thomas, 2000).

Within a tree, the non-conducting heartwood provides a wide and solid central support to the trunk and is surrounded by the water-filled sapwood (275, 293). The vessels in the heartwood frequently become blocked by tyloses. These intrude from adjacent living xylem parenchyma cells, through the vessel pits, and into their lumens (316). The original food reserves and water become progressively withdrawn from the living ray and axial parenchyma cells, their walls often become lignified, and the cells eventually die. Overall, the heartwood becomes drier and frequently becomes filled with gums and resins.

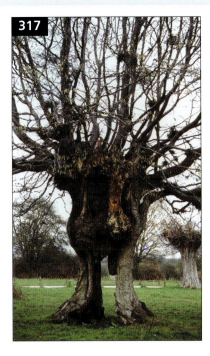

317 The aerial components of an ancient pollard of the broadleaved deciduous *Carpinus betulus* (hornbeam) growing in England. The heartwood centre has rotted away so that the aerial tree appears divided except at the top of its trunk.

318 The hollow heart of an old pollard of *Fagus sylvatica* (beech) growing in Scotland is revealed by the splitting away of part of the trunk. Adventitious roots have grown down from the crown to the ground, to tap nutrients released from the rotten heartwood.

Collectively, the tyloses and such deposits provide at least a partial barrier against the vertical spread of infectious micro-organisms through the wood (see also Chapter 12). It also seems that resins in tracheary elements located at the boundaries between growth rings and in the medullary rays help make resistant barriers to the spread of infection through the wood (Thomas, 2000).

Some trees lose their heartwood as a result of termite assault (estimated to occur in up to half of Australian eucalypts), or fungal and microbial infections (Chapters 8 and 9). However, such trees often survive for many years, as is strikingly demonstrated in various veteran trees (**317**). In such ancient specimens, abundant adventitious roots frequently develop in the upper trunk and grow down its hollow interior, so reclaiming some of the nutrients released from its decayed heartwood (**318**).

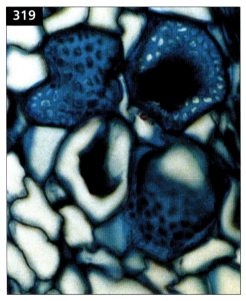

319 TS of *Cucurbita* sp. phloem showing several sieve plates with large pores.

STRUCTURE AND FUNCTION OF SECONDARY PHLOEM (INNER BAST)

The sieve tubes in the phloem of broadleaved species are also specialized. Sieve tubes are formed from several elongated individual cells joined lengthwise, with their end, and frequently side, walls perforated by pores (**319, 320**). These pores vary from several to about 40 µm in diameter, and form simple or compound sieve plates (**319–321**) on the end walls, which partly separate the individual sieve elements (**320, 322**). In functioning phloem (**321, 322**), these pores are generally considered to be open, but in dead tissue they are blocked by plugs of a proteinaceous material and lined by a complex polysaccharide termed callose. The individual elements of a sieve tube have lost their nuclei but, despite appearing empty when viewed under the light microscope, they still

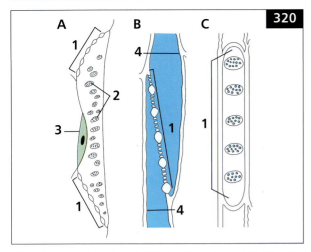

320 Diagrammatic views in LS of a sieve element (**A, B**) with compound sieve plates (1) on the oblique end walls, and sieve areas (2) on side walls. Companion cell (3) and plasmalemma of sieve element (4). (**C**) Face view of sieve plate.

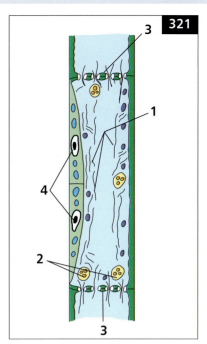

321 Diagram of a mature functional sieve element and companion cell. The former has lost its nucleus, but its plasmalemma is intact and various organelles are still present within its lumen. Proteinaceous fibrils (1), modified plastids (2), sieve plate (3), and companion cell nucleus (4).

contain the other living cellular components (**321**). Sieve tubes are associated with nucleated companion cells with dense cytoplasm. Both elements are intimately involved in the transport of foodstuffs in solution, from the sites of photosynthesis in the foliage to other sites in the tree (**322**). In conifers and tree ferns, companion cells are absent and translocation occurs through the individual, elongated sieve cells, which are provided with numerous small sieve areas on their walls. The newly formed secondary phloem is often only active in translocation for a few months. However, the phloem in some trees, such as species of *Tilia* (lime or linden, **312**), functions for several seasons, while in palms it remains active for the many years of a tree's lifespan (Tomlinson, 2003).

The living axial (longitudinally orientated) parenchyma, and transversely orientated ray parenchyma cells of the secondary phloem provide an important food store. At bud break in deciduous trees (**285**), large quantities of carbohydrates and nitrogenous substances are mobilized from such parenchymatous tissues and transported to support the growth of the expanding leaves. In the tropical tree *Hevea brasiliensis* (rubber tree) the secondary phloem consists of alternating layers of sieve tubes and laticifers, the latter being a series of interconnected latex-secreting cells, which are the

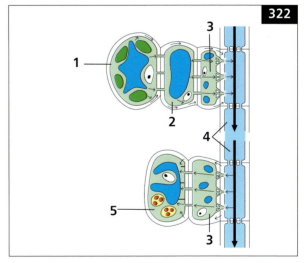

322 Diagram illustrating the translocation in solution of sugars (photosynthesized in the foliage leave chloroplasts), via the phloem sieve tubes in the secondary phloem, to storage tissues elsewhere in the tree. Green photosynthetic cell (1), parenchyma cell enclosing a leaf veinlet (2), companion cell (3), sieve tube element (4), and storage cell of root/shoot (5).

source of commercial rubber. In tapping a tree, a fine half spiral groove is cut into the bark to penetrate to the bast. The latex flows out for several hours and is collected in a cup.

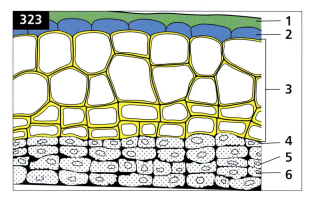

323 Diagram showing the origin of the cork cambium (phellogen) in the outer cortex of a twig, as is common in a broadleaved tree. Cuticle (1), epidermis (2), cork (3), phellogen (4), phellem (5), and cortex (6).

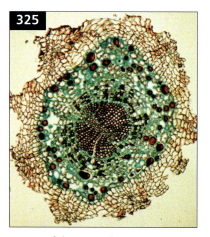

324 TS of a *Sambucus nigra* twig with secondary thickening, showing a lenticel (1). Cork (2), secondary phloem and xylem (3, 4), and primary xylem (5).

STRUCTURE AND FUNCTION OF BARK

DEVELOPMENT OF NORMAL BARK

As secondary thickening progresses, the circumferences of the tree trunk, its branches, and its roots increase (276, 277). On the outside of the vascular cylinder, the original primary cortex and its covering epidermis become stretched. These tissues are replaced, usually in the first year of secondary thickening, by a layer of cork (276D, 277D). The cork develops from the cork cambium (phellogen). In a twig, this typically arises in the cortex (278, 323) but in species of trees such as *Pyrus* and *Quercus suber* (cork oak) it forms in the epidermis. The phellogen also frequently cuts off an internal layer of phelloderm or secondary cortex (324). Cork is also present on tree roots (292, 325). Here, the phellogen forms from the pericyclic parenchymatous tissue situated between the primary and secondary phloem (277B–C).

The cork cells are dead and tightly packed together (278, 325) and, due to the impregnation of their walls with suberin, they are impermeable to both water and air (Chapter 7). However, at the lenticels, or 'breathing' pores, the cork cells are only loosely attached to each other (324, 326) and allow gaseous exchange between the internal tissues of the tree and the external atmosphere. In commercial cultivation of *Quercus suber* (cork oak), the initial bark is first stripped from the trunk to induce the formation of a new phellogen in the innermost secondary phloem. This process is then repeated when each new crop of cork is harvested in approximately 10-year cycles (326).

325 TS of the secondary thickened root of *Ginkgo biloba* showing an extensive outer layer of cork.

326 Upper trunk of *Quercus suber* (commercial cork tree) showing the original bark and surface stripped of cork. Note the growth layers in the cork revealed at the junction of these surfaces.

In smooth-barked trees such as *Fagus sylvatica* (beech, **327**), only a thin layer of secondary phloem forms each year and the initial cork cambium may persist throughout the life of the tree. In the beech phloem, the tips of the radially orientated medullary rays are expanded laterally to form a parenchymatous expansion tissue (**328A**), which accommodates the increasing girth of the tree. In young stems of *Tilia* (lime) species, expansion tissue is also present (**301**) and this initially allows a smooth bark to develop. Later, however, deeper-sited phellogens develop and the bark becomes cracked into shallow plates.

In most trees the secondary phloem develops much more quickly than in *Fagus sylvatica*, and the first-formed phellogens become replaced by new cork cambia. These arise initially from the inner cortex, and subsequently from the older, no longer functional, secondary phloem. In some smooth-barked trees such as species of *Platanus* (plane), *Eucalyptus*, and *Betula* (birch), the increasingly deeper-formed cork cambia develop as partial cylinders. The older bark peels away in sheets (**329**), but these often remain partly adhering to the inner bark (**330**).

In other tree species, the new phellogens frequently develop as a series of overlapping shells to form a compound tissue (rhytidome), in which segments of phelloderm and older secondary phloem become cut off together (**328B**). The scaly bark sloughs segments

327 Cross-cut trunk of *Fagus sylvatica* showing the thin layer of bark (1) with incised initials. Secondary phloem and xylem (2, 3), and vascular cambium (4).

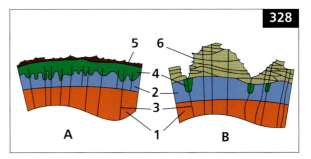

328A–B TS of a smooth-barked specimen of *Fagus* sp. (**A**), and a rough-barked specimen of *Quercus* sp. (**B**). Secondary xylem and phloem (1, 2), rays (3), expansion tissue (4), bark (5), and rhytidome (6).

329 Trunk of *Eucalyptus pauciflora* with various layers of bark flaked from its surface.

330 Trunk of *Betula davurica* showing its bark with very irregular peeling.

off from its surface in an irregular pattern (**331**). In some trees, the secondary phloem contains numerous lignified fibres, which become incorporated into the rhytidome. The bark is consequently very rough and ridged (**275, 314**).

DEVELOPMENT OF WOUND BARK

On many broadleaved trees, shoots commonly sprout from buds buried in the bark of the trunks. These epicormic buds usually represent axillary buds which have remained dormant on the trunk since their formation on the sapling stem many years earlier. These buds grow slowly to keep pace with the expansion of the tree, and sometimes branch prolifically within the bark to create an obvious bur on the trunk. Epicormic buds often become active after the tree is damaged or pollarded, but in species of *Tilia* (lime) and some other trees, even undamaged specimens show profuse growth of new shoots at the base of the trunk (**312, 332**). Adventitious buds may also develop from the callus forming from the inner bark of a logged tree stump or branch (**333**). In conifers, sprouting from buds on the trunk is rare (Thomas, 2000) but does occur in *Sequoia sempervirens* (coastal redwood) and *Araucaria araucana* (monkey puzzle).

331 Trunk of *Pinus sylvestris* showing its irregularly flaked bark.

332 Epicormic buds sprouting in spring from the trunk of *Tilia* x *europaea* (lime) in Scotland.

333 Stump of *Populus* sp. (poplar) growing in Scotland, showing numerous epicormic buds developing in the region of the vascular cambium and also from the bark.

The protective bark of trees is often damaged by lightning strikes, violent gales causing broken boughs (334), browsing animals, and also during branch pruning or tree felling (333; see also Chapter 12). As a result of such damage, the parenchyma and other living cells at the injured surface are killed, and their remnants become impregnated with suberin, a waterproofing waxy material. The lumens of the tracheary elements in the adjacent sapwood also become plugged by gums (Smith, 1986). Bark wounding was investigated experimentally in *Hevea brasiliensis* (rubber) and *Hibiscus rosa-sinensis*.

Longitudinal strips of bark, about 13 mm wide, were excised from their stems (Sharples and Gunnery, 1933). A parenchymatous callus developed on the exposed sapwood surface, and new cork and vascular cambia formed within the callus – in continuity with those of the intact stem – until eventually joining together, with the wounds healing over within a couple of months. Similar marginal wound scar healing (335A), and the development of nodular callus (335B) from the exposed sapwood medullary ray parenchyma, occur in *Quercus petraea* (sessile oak, Bowes, 1999).

334 Massive bough of *Quercus petraea* (sessile oak) broken off during a violent storm in Scotland.

335A–B Large wound on the trunk of *Quercus petraea* (sessile oak) growing in Scotland. The wound (A) is approx 35 x 18 cm, and a prominent ridge of wound cork has formed at both its sides. Nodular callus (lower right) occurs over part of the exposed sapwood. B, detail of the nodular callus formed from the xylem medullary ray parenchyma. The rays show on the wood surface as narrow vertically elongated structures.

A frequent wounding response, involving the formation of new bark, is in the healing of a branch scar after tree pruning (336). In nature this occurs on various broadleaved and coniferous tree species after self-pruning. In the latter case an abscission zone develops at the base of a branch, and internal to this zone a thin layer of cork is formed prior to branch abscission (Bell, 1991). Eventually, even quite large scars will often heal over. However, this is not possible if the branch does not fall off cleanly and a persistent decayed core remains (337).

Another common example of the wounding reaction in trees is in the natural grafting of roots where they are in intimate contact in the soil, or in the aerial roots of various species of *Ficus* (338). This grafting also frequently occurs on the woody climbing stems of *Hedera helix* (common ivy, Milner, 1932), and sometimes on young branches of a tree growing in a protected site, or in saplings planted very close together to form a hedge (339, 340).

336 Pruning wound on the trunk of *Aesculus hippocastanum* (horse chestnut) growing in Scotland. Note how rings of wound callus have partly occluded the original cut surface of the small side branch.

337 Trunk of *Quercus petraea* (sessile oak) growing in Scotland, showing the dead core of a side branch with a ridge of wound cork at its origin from the trunk.

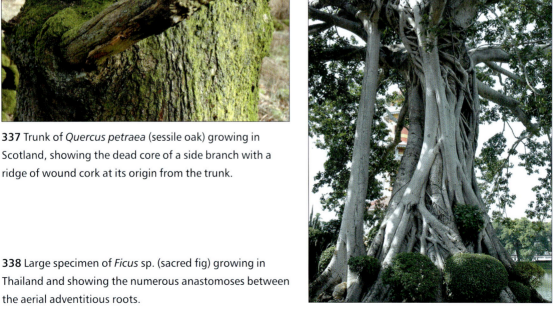

338 Large specimen of *Ficus* sp. (sacred fig) growing in Thailand and showing the numerous anastomoses between the aerial adventitious roots.

339 Natural grafting between a large stem and smaller side branch of *Fagus sylvatica* (beech) growing in a sheltered area in Scotland.

340 Part of the famous 'hedge' of *Fagus sylvatica* (beech) at Meikleour, Scotland, which was planted in the mid-18th century. Note how several trunks of the now tall trees have fused laterally, due to being planted so close to each other.

SECTION 4 TREE PATHOLOGY

CHAPTER 7

The role of cell-wall polymers in disease resistance in woody plants

Christopher T Brett

INTRODUCTION

The plant cell wall is a highly effective barrier to penetration by potential pathogens, whether they are micro-organisms or higher organisms. Each species of tree exhibits 'non-host resistance' against the great majority of potential pathogens. This means that all varieties of that species are resistant to all strains of the potentially pathogenic species. This broad resistance is due, among other things, to the structural strength of the cell wall, together with its tightly-knit structure, through which only very narrow pores penetrate.

In the young and active tissues of trees and other plants (**341, 342**), the mechanical strength of their cell walls is provided by the cellulose–hemicellulose network (**343**), while the pore size is controlled by the pectin network. The two networks act together to provide a structure which is mechanically strong and resistant to penetration, yet retains the ability to stretch under controlled conditions and permit cell expansion. In the lignified fibres and tracheary elements of the vascular system (**344**), the mechanical strength of their cell walls is greatly increased, while porosity is reduced further by the loss of water from the cell walls.

In addition to these passive roles in resistance to pathogens, cell walls also participate in the active resistance mechanisms by which trees respond to pathogen attack. These mechanisms involve the formation of fragments of cell-wall molecules by the action of cell-wall-degrading enzymes. These fragments then act as inducers, named 'elicitors', of active defence mechanisms in the plant. Some of these defence mechanisms in turn involve modification of the plant cell wall.

This chapter will review the molecular structure of the plant cell wall, and then discuss the various roles of the wall polymers in plant defence. While the principles involved are common to all types of plants, the application to trees and woody plants will be emphasized where appropriate.

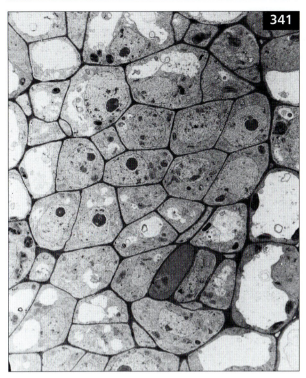

341 Low-resolution electron micrograph showing procambial tissue in transverse section of a young stem of *Glechoma hederacea* (ground ivy). (Photo copyright of Bryan Bowes.)

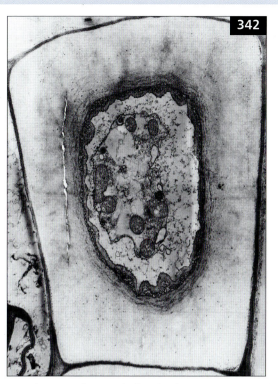

342 Electron micrograph showing in transverse section the thick cellulosic wall of a still-developing fibre in *Linum usitatissimum* (flax). (Photo copyright of Bryan Bowes.)

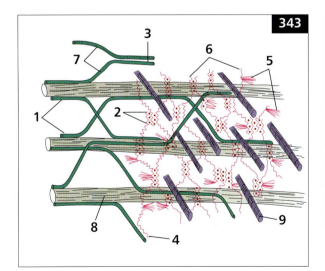

343 Model of the molecular structure of the primary plant cell wall. 1, hemicelluloses hydrogen-bonded to microfibrils; 2, polygalacturonic acid chains ionically cross-linked by calcium ions; 3, possible hydrogen-bonding between hemicellulose molecules in the cell-wall matrix; 4, some covalent cross-linking may occur between xyloglucan and RG-I (5); 6, pectic polysaccharides; 7, hemicelluloses; 8, cellulose microfibril; 9, protein.

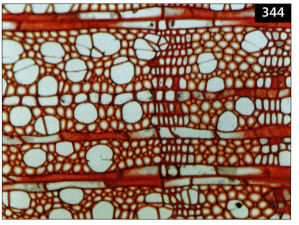

344 Light micrograph showing a transverse section of the wood (secondary xylem) of the broadleaved tree *Magnolia grandiflora*. Note the thinner lignified secondary walls of the large-diameter vessels, and thicker walls of the smaller tracheids and fibres. (Photo copyright of Bryan Bowes.)

THE MOLECULAR STRUCTURE OF CELL-WALL POLYMERS

THE CELLULOSE–HEMICELLULOSE NETWORK

Cellulose consists of long chains of 1,4-β-glucans, with 30 or more chains being aligned in parallel to form a microfibril (345). These microfibrils are held together by numerous hydrogen bonds, resulting in a structure that is crystalline in its core, and partially crystalline at the exterior. These microfibrils are themselves generally aligned in parallel in each layer of the wall (346), with the direction of alignment varying from one wall layer to the next. Since the microfibrils provide a high degree of tensile strength along their longitudinal axes, this arrangement gives all-round strength similar to that found, on a different scale, in plywood.

However, the microfibrils do not bind laterally to one another, so further components are required to hold them in place (347). These components are part of the cell-wall matrix, which surrounds the microfibrils. The main matrix polymers which hold the microfibrils in place are the hemicelluloses. These are polysaccharides which have a backbone with a secondary structure similar to cellulose, and are able to hydrogen-bond to the microfibrils. By bonding to two neighbouring microfibrils, and forming a cross-bridge, or tether, between them, the hemicelluloses hold the microfibrils in place. The strength of the cross-bridges is thought to be regulated by proteins called 'expansins'. These proteins are able to break the hydrogen bonds between hemicelluloses and

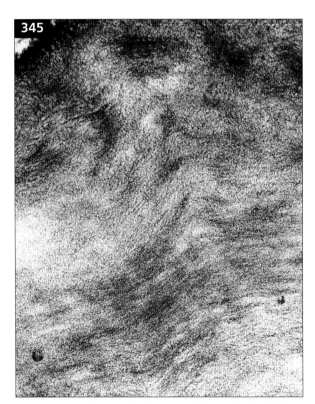

345 Electron micrograph showing evidence of microfibrils in a glancing section of the primary wall of a higher plant. (Photo copyright of Bryan Bowes.)

346 Electron micrograph of the plant primary cell wall/plasmalemma interface (as seen in a freeze-fractured specimen). Note the parallel arrangement of the microfibrils in the wall. (Photo copyright of Bryan Bowes.)

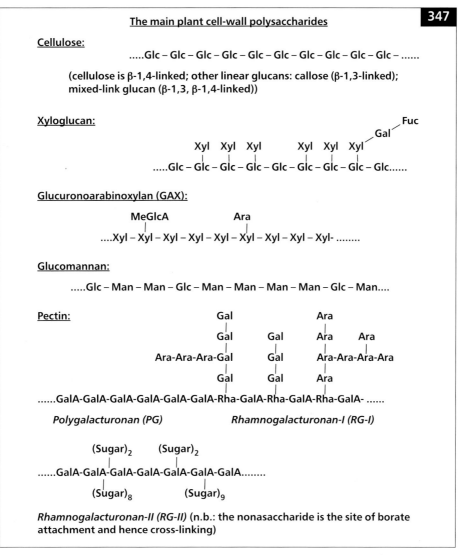

347 The main plant cell-wall polysaccharides.

cellulose, thus weakening the tethers between microfibrils. Expansins are thought to have a key role in weakening the hemicellulose cross-bridges sufficiently to allow cell expansion.

In the primary walls of dicots, including broadleaved trees, the main hemicellulose involved in tethering microfibrils is xyloglucan. Glucuronoarabinoxylans are also present in smaller amounts. In the secondary wall (that which may subsequently be formed by the protoplast at the end of its expansion, as in fibres and tracheary elements), glucuronoxylans are the major hemicellulose, and are thought to act as tethers in the same way. In the secondary walls of gymnosperms, glucomannans fulfil the same function.

THE PECTIN NETWORK

The cell-wall matrix contains an additional network, the pectin network, which is present in the spaces between the microfibrils and surrounds the cellulose–hemicellulose network, but has relatively few covalent or non-covalent bonds with it (**343**). The pectin network is especially important in the middle lamella, the junction zone between cells, which has relatively little cellulose. In contrast, the secondary wall has very little pectin. Pectin consists of galacturonic-acid-rich polymers, which are highly hydrated due to the negatively-charged galacturonate residues (**347**). The simplest pectin domain is polygalacturonic acid (PGA, also known as homogalacturonic acid, or HGA). This consists of linear chains of galacturonic acid residues, some of which have methyl groups esterified to the galacturonate carboxyl groups.

The other major pectin domain is rhamno-galacturonan-I (RG-I), which contains a backbone of alternating galacturonic acid and rhamnose residues, with neutral side-chains containing arabinose and galactose attached to some of the rhamnose residues. A further, minor component is rhamnogalacturonan-II (RG-II), which has a backbone of PGA but also contains four complex side-chains containing rhamnose, galactose, and a variety of rare sugars such as aceric acid and apiose. Another pectin domain that is sometimes present is xylogalacturonan (XGA), which has a PGA backbone but also has side-chains containing xylose residues. How the different pectin domains are linked together covalently is not yet clear. It is known, however, that there are numerous cross-links between pectin domains, especially through ionic bonds involving calcium ions which form ionic bridges between galacturonate groups on neighbouring PGA domains. RG-II domains can also form cross-links with each other, through borate ester links between apiose residues. There is good evidence that the pectin network controls the porosity of primary cell walls, and it probably also controls intercellular adhesion due to its predominance in the middle lamella.

HYDROPHOBIC NETWORKS

The polysaccharide-based networks which form the major part of most cell walls are relatively hydrophilic. These networks are well-suited to plants growing in an aqueous environment. However, trees and other land plants require support against gravity and are subjected to desiccation stresses. The hydrophobic networks provide protection against desiccation, and support against mechanical stresses. They also provide highly effective barriers to invasion by micro-organisms.

The outer surfaces of the young stem and its leaves (but not its roots) are covered by a cuticle (348). The outermost layer of the cuticle consists of waxes, which are hydrocarbons of 18–24 carbons in length. These provide a hydrophobic layer which is not readily wetted by water droplets. This in itself is a defence against infection, since micro-organisms are often transported between plants by water droplets. Beneath the wax layer is a layer of cutin, which forms the bulk of the cuticle. Cutin is made up of C16 and C18 hydroxylated fatty acids, linked together by ester bonds to form an insoluble polymeric network (349). This layer is also hydrophobic, and because of the ester cross-links between the polymers, forms an effective defensive

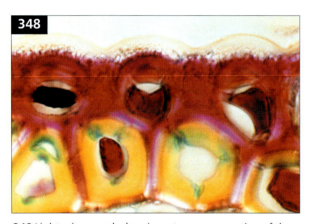

348 Light micrograph showing a transverse section of the outer surface of the leaf of *Pinus monophylla* (nut pine). Note the deep cuticular covering of the thick-walled (red-coloured) epidermal cells. (Photo copyright of Bryan Bowes.)

349

Ester linkage between C$_{16}$ molecules

Ester linkage to other hydroxyl groups

Ester linkage to phenolic acid

349 The structure of cutin.

350 The structure of suberin.

351 The outer surface of the bark of a large trunk of *Psuedotsuga menziesii* (Douglas fir). (Photo copyright of Bryan Bowes.)

barrier against penetration of the tissue by potential pathogens.

The corresponding surface covering of the older tree and its roots contains suberins, which are made up of cross-linked, hydroxylated fatty acids of 20 or more carbon atoms in length, and which also contain some cross-linked phenolic groups (**350**). Suberin is a major component of bark and cork (**351**), and is also found in internal hydrophobic walls in the Casparian strips of the endodermis (**352**).

The main role of both cutin and suberin is in limiting the movement of water. The other major hydrophobic polymer is lignin, which is not only highly hydrophobic but also has a major structural role in the xylem elements and fibre cells (**344**). Lignin

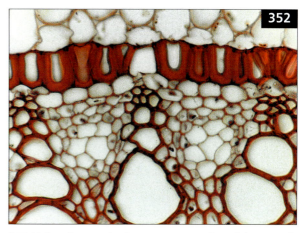

352 Light micrograph of a transverse section of the root of *Iris* showing a conspicuous single-layered, thickened endodermis, while internally numerous tracheary elements are visible. (Photo copyright of Bryan Bowes.)

is a highly cross-linked polymer of phenylpropanoid units (353), generated by the spontaneous free-radical coupling of the electron-deficient radicals produced by the action of oxidative enzymes on the phenylpropanoid alcohol precursors. The main enzyme involved is peroxidase, which uses hydrogen peroxide as the oxidizing agent. Laccase, which uses oxygen, may also be involved. By displacing water in these cells, lignin prevents relative movement of the polysaccharides, and eliminates any possibility of cell extension. This results in an extremely strong cell wall, which can resist the mechanical stresses imposed by gravity and wind, and also those resulting from the hydraulic forces generated in the water-conducting elements of the xylem.

Phenylpropanoid units are also found in small amounts in the young, primary walls of growing tissues. Here they are esterified to matrix polysaccharides, and can cross-link by oxidative coupling, as in lignin polymerization. These cross-links strengthen the wall, decreasing its extensibility, and at the onset of lignification may act as nucleation sites for lignin polymerization.

While the primary physiological roles of these hydrophobic polymers are in the control of water movement and the provision of mechanical support, they all have the additional property of providing defence against pathogens. This will become clear in later sections of this chapter.

THE STRUCTURAL PROTEIN NETWORK

The cell wall contains a wide range of proteins. Many of these are enzymes, involved in cell-wall metabolism and/or defence reactions. However, there are several classes of wall proteins, the function of which is thought to be primarily structural. The most important of these are the hydroxyproline-rich glycoproteins, or HRGPs, also known as extensins. As the name implies, these contain hydroxyproline, an amino acid absent from most proteins. They also contain tyrosine residues, which are capable of cross-linking the polypeptide chains by the formation of isodityrosine bonds between two tyrosine side-chains. Since HRGPs rapidly form a highly insoluble network in the cell wall, it is thought that the isodityrosine cross-links are intermolecular rather than intramolecular, although

353 Structure of lignin.

this has yet to be demonstrated conclusively. The cross-links are formed by the action of peroxidase and hydrogen peroxide, an oxidative reaction similar to that involved in cross-linking ferulic acid residues attached to matrix polysaccharides.

THE CELL WALL AS A PHYSICAL BARRIER TO INFECTION

Even in a young, hydrated cell wall, the aqueous pores are sufficiently small to prevent the diffusion through them of molecules >100 kDa. Hence the cell wall forms an effective barrier to the penetration of plant tissues by bacteria, fungi, and larger organisms. The hydrophobic components of cell walls (cutin, suberin, and lignin; 344, 348, 352) are even more effective in preventing

microbial penetration. For effective colonization of plant tissues, micro-organisms must first penetrate through the surface of the tree or other plant. Then, to obtain access to the cytoplasm of plant cells, micro-organisms must penetrate the cell wall itself. Several strategies have evolved for each of these steps.

To pass through the surface layers of plant organs, micro-organisms can use natural openings in the plant surface, especially the stomata on the surface of leaves and young stems (354). Both bacteria and fungi can enter through stomata (355), as can some nematodes. The corky, suberized surfaces of trees and other woody plants contain openings called lenticels, which allow the movement of air into the underlying tissues (356, 357). These

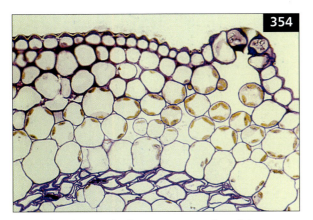

354 Light micrograph of a transverse section of the young stem of *Phaseolus vulgaris* (French bean) showing a prominent stoma in the epidermis. (Photo copyright of Bryan Bowes.)

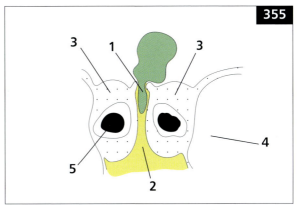

355 Diagram of a fungal hypha penetrating into a leaf via a stoma on its epidermis. 1, tip of fungal hypha; 2, stomatal cavity within leaf; 3, thickened walls of guard cells; 4, large vacuolated subsidiary cells to stoma; 5, nuclei of guard cells.

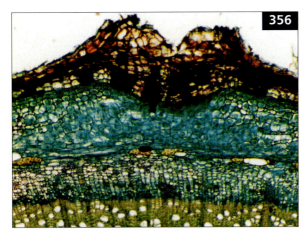

356 Light micrograph showing a transverse section through a lenticel in the bark of *Sambucus nigra* (elderberry). (Photo copyright of Bryan Bowes.)

357 Trunk of *Betula pubescens* (downy birch) showing numerous transversely elongated lenticels. (Photo copyright of Bryan Bowes.)

lenticels are also possible points of entry for pathogens, although pathogens that can penetrate through lenticels are usually more effective when they penetrate through wounds. Many micro-organisms can penetrate through wounds, where the normal surface layers of an organ have been broken, exposing the cell walls of the interior of the tissue. Wounds are especially important for bacterial infection; for instance in trees, the causative agent of crown gall disease (*Agrobacterium tumefaciens*) requires quite fresh wounds for penetration. Viruses may also penetrate through wounds, although where insect vectors are involved, the wounds are made by direct penetration of the plant surface by the insect.

Many fungi are able to penetrate young, intact plant surfaces, without making use of wounds or natural openings. They do this partly by degrading the cutin layer, by secreting cutinases. These enzymes are able to hydrolyse the fatty acid ester bonds which cross-link the hydroxylated fatty acid components of cutin. This weakens the cutin layer sufficiently to allow penetration of a fungal hypha through the cutin. Another strategy is the generation of sufficient mechanical force to force a penetration peg through the cuticle (358). Some fungi can achieve this by forming appressoria which bind very strongly to the surface of the cuticle, and then generating sufficient osmotic pressure to achieve penetration over a small area of the cuticle surface. It is likely that the point at which penetration occurs is weakened by localized secretion of cutinases and cell-wall-degrading enzymes from the tip of the advancing penetration peg.

Once through the outer, hydrophobic layer of the plant surface, further penetration can be achieved by secretion of enzymes which degrade the cell wall. Many fungi and bacteria secrete pectinases. These include polygalacturonases, which degrade PGA hydrolytically, at points where the degree of methylation of the galacturonic acid carboxyl groups is low. Other pectinases include pectate lyases and pectin lyases, which catalyse elimination reactions that cleave pectate (demethylated pectin) and methylated pectin, respectively. Pectin methyl esterases (PMEs) remove the methyl ester groups, thus facilitating the action of those pectinases that act on demethylated pectin. An additional consequence of PME action is to lower the wall pH by generating free carboxyl groups in PGA, and

thus accelerating pectin degradation by those pectinases which have acid pH optima. The RG-I backbone is cleaved by rhamnogalacturonases, while the side-chains are degraded by β-galactanases and α-arabinanases.

Pectin degradation is sufficient by itself to break down or greatly weaken the middle lamella, which is composed mostly of pectin. Hence, secretion of pectinases permits the movement of fungal hyphae between cells, allowing penetration deep into the tissue and at the same time greatly weakening the overall strength of the plant tissue. This strategy allows fungi to absorb those nutrients which are present in the apoplast (the cell walls and intercellular regions), including sucrose and some amino acids.

Pectin degradation alone is not normally sufficient to allow degradation of the primary wall. Many fungi secrete a further range of enzymes to degrade hemicelluloses. These include xyloglucanases, which are sometimes classified as

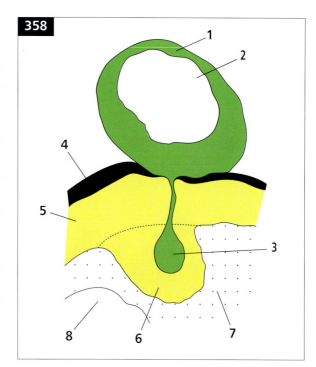

358 Diagram showing papilla formation at site of fungal penetration through a plant cell wall. 1, fungal hyphal cytoplasm; 2, hyphal vacuole; 3, fungal tip encased in cell-wall papilla of host plant; 4, cuticle of host epidermis; 5, normal cell wall of host cell; 6, multilayered papilla in host cell wall; 7, cytoplasm of host plant cell; 8, vacuole of host cell.

cellulases or 1,4-β-glucanases, since they break the 1,4-β-glucan backbone of xyloglucan. Many xylanases are also produced by fungi, which are capable of degrading the arabinoglucuronoxylan of the primary wall and the glucuronoxylan of the secondary wall. Another enzyme secreted by fungi is 1,3-β-glucanase, which degrades the 1,3-β-glucan, callose. Callose is not normally part of the primary cell wall of healthy higher plant cells, except in certain specialized tissues such as phloem sieve tubes, pollen tubes, and developing pollen grains. However, it is produced by most plant cells as a wound response. In the case of fungal infection, the presence of a fungal hypha on the surface of a plant cell often triggers the formation of a new layer of cell-wall material on the inside surface of the normal primary wall, called a papilla, at the point at which attempted penetration of the primary wall by the fungus is occurring (358). This papilla is constructed mainly of callose. Hence, those fungi that secrete 1,3-β-glucanase may do so in order to break down this callose layer produced in response to the fungal attack.

Many micro-organisms secrete enzymes classified as cellulases. It is not always clear whether the substrate of these enzymes is cellulose itself, or xyloglucan, which has the same backbone structure. If cellulose is in fact degraded, this is likely to cause a significant weakening of the wall and facilitate its penetration.

Lignin forms an even more effective barrier to microbial penetration than the polysaccharide components of the wall. Because it is produced by the non-enzymic coupling of phenolic free radicals, many different types of bond form between the neighbouring phenylpropanoid units of lignin. These include carbon–carbon bonds and ether links, both of which are extremely stable chemically, and resistant to enzymic degradation. In addition, the fact that lignification is accompanied by the loss of most of the water from the cell wall means that diffusion of enzymes into a lignified cell wall is almost impossible. Most micro-organisms are therefore unable to degrade lignified woody tissue. The exceptions are the wood-rotting fungi and a small number of bacteria. These micro-organisms secrete oxidative enzymes, which produce powerful oxidants capable of breaking open the stable bonds that are resistant to normal enzymic attack. The main wood-rotting fungi are the white rot fungi, a group of basidiomycetes which are responsible for recycling the massive amounts of lignin generated by trees. The fungal degradation of lignin prevents the excessive accumulation of dead wood in the biosphere. Apart from its lack of resistance to these few wood-rotting organisms, lignin is a highly effective barrier to pathogenic attack. This means that normal lignified tissues have a natural defence against most potential pathogens. For instance, elms resistant to Dutch elm disease have a high proportion of thick-walled, lignified xylem vessels in their vascular tissues. In addition to this, lignin is actively laid down by the plant at points of microbial attack, to prevent further penetration into the tissue. This is one part of the active defence mechanisms which higher plants develop in response to potential pathogens (see below).

Suberin also plays an important role in the resistance of woody plants to pathogens. In a study of the role of cell-wall polymers in the resistance of *Musa* (banana) to burrowing nematodes, resistant cultivars were found to have increased suberin in their endodermis, and the nematodes could not penetrate further than the cortical layers of the root. The amount of callose in the cortical cells was also increased in these resistant cultivars (Valette *et al.*, 1997).

ACTIVE DEFENCE MECHANISMS INVOLVING THE CELL WALL

In addition to the passive defence offered by the cell wall, trees and other higher plants possess a series of active response mechanisms that have evolved to prevent the spread of infection by micro-organisms. Many of these mechanisms involve the cell wall. Some of them bring about reinforcement of the cell wall, making it more difficult for infection to spread through the plant tissues. Others involve cell-wall fragments acting as molecular signals, which trigger defence mechanisms. These fragments may be generated by the action of fungal enzymes on cell-wall polymers. Most of the active defence mechanisms are initiated by the action of these molecular signals, also known as 'elicitors', in stimulating the formation of 'pathogenesis-related proteins', or PRPs. PRPs include enzymes, structural proteins, and perhaps also receptors capable of recognizing the presence of elicitors and initiating appropriate defence responses.

CELL-WALL REINFORCEMENT IN RESPONSE TO PATHOGEN INFECTION

One class of PRPs is the extensins, or HRGPs. In general, these structural proteins make the wall stronger, and less easily penetrated by micro-organisms. This is probably because they are rapidly cross-linked by peroxidase, which is also a PRP, i.e. its formation is stimulated in response to infection. Since hydrogen peroxide concentrations also rise in response to infection, all the reagents required for forming the cross-linked protein network are present.

The presence of elevated levels of peroxidase and hydrogen peroxide also brings about increased cross-linking of pectins, through oxidative coupling of the ferulic acid residues attached to the RG-I side-chains. While ferulic acid is usually a very minor component of pectin, there are some plants which contain larger amounts of it, in which case pectin cross-linking is likely to bring about significant strengthening of the cell wall. It is also thought that coupled ferulic acid residues act as initiation points for lignin biosynthesis. If so, then even a low level of ferulic acid dimerization may bring about a major effect on wall properties.

Lignification is the most dramatic change in cell-wall structure that may occur in response to infection. For instance, varieties of *Musa* (banana) resistant to *Fusarium oxysporum* have been found to produce large quantities of lignin and other phenolics in response to elicitors prepared from the fungus, while susceptible varieties produced much lower amounts of phenolics (De Ascenao and Dubery, 2000). When lignification occurs, the cell wall becomes impenetrable to most micro-organisms, and this is a highly effective way of limiting the spread of infection. Lignification also results in the death of the cells concerned, so it is one part of a well-recognized defence strategy, the 'hypersensitive response'. This is the initiation of cell death in the infected area, accompanied by lignification; effectively sacrificing the cells around the point of infection in order to save the rest of the plant from further damage.

Suberin deposition is also an important feature of defence responses in trees. For instance, when resistant varieties of *Cupressus* (cypress) are infected with the fungus *Seiridium cardinale*, the growth of the fungus outwards from the point of infection on the tree is prevented by the formation of several layers of suberized cell walls. Susceptible varieties produce a much thinner layer of suberized cells, which is insufficient to prevent the further growth of the fungus.

Another defence mechanism involves degradation, rather than strengthening, of the cell walls. This is the abscission response, which can bring about the shedding of infected parts of leaves from the rest of the leaf. This response occurs in young *Prunus* leaves infected by fungi, bacteria, or viruses. The middle lamella between the cells at the abscission layer is rapidly and selectively degraded by pectinases, causing a loss of intercellular adhesion, and subsequent separation of the cell layers at this point.

DEFENCE MECHANISMS INVOLVING INHIBITORS OF CELL-WALL-DEGRADING ENZYMES

An additional strategy employed by higher plants as a defence against potential pathogens is the production of enzyme inhibitors. A frequent response to infection involves the production of pectinase inhibitors. Pectin methylesterase inhibitors are also frequently produced, together with proteinase inhibitors, which may interfere with the breakdown of both cell-wall and cytoplasmic proteins.

CELL-WALL-DERIVED MOLECULAR SIGNALS INVOLVED IN INITIATION OF DEFENCE RESPONSES

In order to penetrate plant tissues, fungi and bacteria must degrade cell-wall polymers. They do this by secreting enzymes which break the polysaccharide chains. This results in the formation of polysaccharide fragments ('oligosaccharins'), some of which are small enough to diffuse rapidly through the wall. Higher plants recognize these fragments as signals which indicate pathogenic attack, and use them to trigger defence reactions (359).

The best-known of these cell-wall-derived molecular signals, or elicitors, are the oligo-galacturonides formed by the degradation of polygalacturonic acid by polygalacturonase. These are oligosaccharides containing up to 20 galacturonic acid residues. The exact size required for optimal activity varies from plant to plant and in different tissues. It also seems that optimal activity sometimes requires oligogalacturonide dimerization, by the

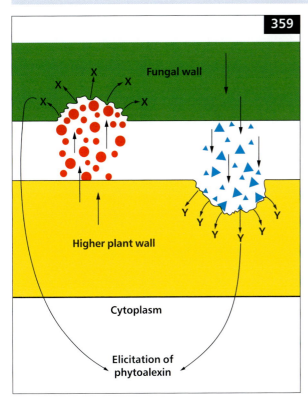

359 Formation of oligosaccharin elicitors at the interface between a higher plant cell wall (yellow) and fungal hyphal wall (green). Red circles, higher plant enzymes; blue triangles, fungal enzymes. X, elicitor derived from fungal wall; Y, elicitor derived from higher plant cell wall.

formation of ionic cross-links involving calcium ions, which bind to the negatively-charged carboxylate groups on both oligosaccharides. However, in other cases, such ionic dimerization is not needed. Oligogalacturonides can elicit a wide range of different defence reactions. They were originally discovered as elicitors of phytoalexins, anti-microbial toxins produced by higher plant cells in response to infection. However, they can also induce many other responses, including lignification and the hypersensitive response.

Other oligosaccharide elicitors are derived not from the host cell wall, but from the cell walls of invading fungi. Plant surfaces contain small amounts of 1,3-β-glucanases and chitinases, and the levels of both enzymes increase during infection, i.e. they are PRPs. Their substrates are two polysaccharides found in fungal cell walls, 1,3-β-glucans and chitin. Part of the antifungal

activity of these enzymes may be due to their effect on weakening the fungal cell wall. However, a more important effect is almost certainly due to the activity of the enzyme products as elicitors. The glucan and chitin fragments produced by the action of the enzymes diffuse back into the host cells, and there act as elicitors of a range of defence responses. In some cases, the oligogalacturonide and glucan elicitors have been found to act synergistically, i.e. the combined effect of the two together is greater than the sum of the effects of each acting separately.

FUTURE DEVELOPMENTS

The cell walls of trees and other higher plants are involved in pathogen defence, both as a passive barrier to pathogen penetration and as a key component of active defence mechanisms triggered by pathogenic attack. To date, the research that has established these principles has involved microscopic and biochemical analysis of cell walls and their interaction with pathogens. This area of research is now being given a new stimulus by the application of molecular genetic techniques. Two strategies seem particularly promising.

First, the complete sequencing of plant genomes makes it possible to study the full range of genes for which transcription is affected by pathogen attack. *Populus trichocarpus* (western balsam poplar) has been selected as a model tree species for the molecular biology of woody plants, and is expected to be the first tree species to have its genome sequenced (Brunner *et al.*, 2004). A microarray study of this species (Smith *et al.*, 2004) has already revealed large numbers of genes, the transcription of which responds to wounding and/or infection by the poplar mosaic virus (PopMV). This work indicates that many of the responses to viral infection may be general responses to wounding, rather than specific responses to viral attack.

Secondly, genetic analysis permits very precise changes in wall structure to be detected, and makes it possible to determine the effects of these changes on responses to pathogens. This approach has so far been applied mainly to the model herbaceous plant, *Arabidopsis thaliana* (thale cress; Vorwerk *et al.*, 2004). However, the availability of the full genetic sequence of poplar, and subsequently other species of trees, will permit similar studies to be undertaken in woody plants.

CHAPTER 8

Microbial and viral pathogens, and plant parasites of plantation and forest trees

Stephen Woodward

INTRODUCTION

Disease leads to harmful changes in the appearance of a tree compared with a healthy individual, altering its growth, reducing amenity value and yields, or even killing it. Although symptoms may be striking, the diagnosis of individual diseases requires a broad knowledge of trees and of the different disease categories affecting them. Factors other than disease, such as poor environmental conditions and nutritional limitations, can also result in abnormal growth, while careless use of herbicides or de-icing salts can also cause damage.

Disease symptoms are visible manifestations of the interactions between the disease-causing agent (the pathogen) and the host tree. These may include changes in leaf or shoot colour, premature leaf abscission, distorted or reduced growth, stunting, wilting, dieback, and decay. In addition to these effects on growth, fruiting bodies of fungal pathogens may be observed on or near the affected tree. Diseases may be caused by viruses, bacteria (including phytoplasma), fungi, and parasitic higher plants.

Historically, there was a tendency to focus on plant diseases only when they threatened human needs in some way, impacting on yields of food crops or other raw materials. It is becoming increasingly recognized, however, that diseases are of fundamental importance in plant communities, with major roles in succession in natural ecosystems. Diseases may act as natural thinning agents in forests, reducing the fecundity of or eliminating poorly adapted host genotypes from the breeding population, and could, therefore, be considered to be a driving force in the evolution of the ecosystem.

This chapter focuses on the different types of disease affecting trees. Some disease-causing organisms are not confined to a single part of a tree and may affect several different tissues. Many different organisms can cause similar types of disease symptoms, but a selection of those with high economic and ecological importance in forests of different parts of the world is described in detail below. To some extent, the diseases chosen reflect those for which a reasonable literature exists, and this fact reflects the human focus on pathogens which have caused significant economic damage.

TREE ROOT DISEASES

Damage to the roots is usually recognized by the appearance of symptoms throughout the whole crown; in some cases, the tree may die for no immediately apparent reason. Without performing appropriate tests, it can be difficult to distinguish disease from abiotic disorders: yellowing of the leaves throughout the crown can result from reduced nutrient and water availability as well as from infections.

DISEASES OF PRIMARY ROOTS

A large number of fungi and fungi-like organisms may cause disease on primary roots lacking extensive secondary thickening. Species of *Phytophthora* cause extremely serious diseases in many areas of the world (*Table 4*, 360–369); they are not fungi, but are more closely related to certain marine algae. Other pathogens, including species of *Pythium*, *Fusarium*, and *Rhizoctonia* agg., also attack fine-root systems.

PHYTOPHTHORA CINNAMOMI: PHYTOPHTHORA ROOT ROT

Many *Phytophthora* diseases affect trees, but not all are root pathogens. *Phytophthora cinnamomi*, however, is of particular significance on a wide range of woody plants, both gymnosperms and angiosperms, in many areas of the world, including North America, Europe, and Australia.

Symptoms and disease cycle

Asexual and sexual spores (362, 363) resistant to dry and cool environmental conditions are of vital importance in the survival of *Phytophthora* spp. In wet soils, these spores germinate, releasing motile spores (zoospores) which infect fine roots and grow into the vascular cambium of the secondary roots. Symptoms are typical of severe loss of root function: small, chlorotic foliage, dieback, and death (360, 361).

Importance

P. cinnamomi was estimated in the 1970s to affect some 282,000 ha of *Eucalyptus marginata* (jarrah) forests in Western Australia and to be increasing by 20,000 ha each year. Jarrah is a prime timber species and this disease threatens regional forest industries. The pathogen also attacks other species of *Eucalyptus* and many of the woody understorey species (especially *Banksia*), leaving a severely depleted ecosystem. In the state of Victoria, *Eucalyptus sieberi* and *E. globidea* forest is seriously affected by *P. cinnamomi*. As in Western Australia, disease incidence increased in the 1950s when power line and road construction in the forests intensified. This ecological change threatens the stability of water catchment areas.

In the southern USA, *Pinus echinata* (shortleaf pine) planted on severely degraded and nutrient-limited former cotton-growing lands developed little-leaf disease following *P. cinnamomi* infection.

Management

The disease is extremely difficult to control because of the resilient spores present in the soil. General strategies for managing *Phytophthora* include providing good drainage, increasing tree vigour by fertilizer application, and conversion to less-susceptible species. In Australia, a programme of selection of *E. marginata* for resistance to *P. cinnamomi* is in progress. Trials of chemical control using phosphite injection into tree trunks have shown some promise, but the method used is both costly and time consuming.

Further 'new' species of *Phytophthora* have caused major damage to trees in the last 10 years, good examples being *P. ramorum* (see pp. 143 and 154) and *P. alni* in northern Europe, which attacks riparian alder trees resulting in host death (365).

Table 4 Species of *Phytophthora* of particular economic and environmental importance for trees. Several previously unrecognized species have been reported in the last decade, and are referred to as 'recently emerged'

Species	Locality	Principal hosts	Damage	Figures	Comments
Phytophthora cinnamomi	World-wide	Very wide host range	Reduced growth; death. Major ecosystem changes	360–363	Serious in Western Australia, Victoria State, Southern USA
Phytophthora cambivora	Europe	Wide host range	Root death; reduced host growth; death	364	Ink disease of *Castanea*
Phytophthora alni	Northern Europe	*Alnus* spp.	Small leaves; thin crown; dieback; death	365	Recently emerged in Europe; possible hybrid between *P. cambivora* and species close to *P. fragariae*
Phytophthora lateralis	Coastal northern California and southern Oregon	*Chamaecyparis lawsoniana*	Host death. Major ecosystem changes	366, 367	Local economies based on timber seriously affected
Phytophthora ramorum	Pacific north-west Europe	Wide host range	Rapid death of foliage and shoots	368, 369	Recently emerged in North America; Sudden death of oak. Recently found in Europe
Phytophthora quercina	Northern Europe	*Quercus, Fagus*	Root death, small leaves; reduced host growth	–	Recently emerged in Europe
Phytophthora cactorum	North America, Europe	Wide host range, angiosperms	Collar rot; attacks roots, twigs and fruits; basal and bleeding cankers on trunks	–	Widespread pathogen causing collar rot of orchard and ornamental trees

DISEASES OF SECONDARY ROOTS

Fungi degrading secondarily thickened tissues produce specialized oxidative enzymes for growth in this recalcitrant substrate (see also Chapter 7). Most of the fungi capable of degrading lignified tissues are not pathogens in the true sense, but exist saprotrophically in dead woody tissues, causing decay. A limited number of these species can enter the roots of living trees causing growth reductions and even death, and may, therefore, be described as pathogens. This section focuses on two groups of organisms causing disease on secondary root systems, *Armillaria* (honey fungus, 370–375) and *Heterobasidion* (fomes, 376–378). However, several other species are of importance (*Table 5*, 379–383).

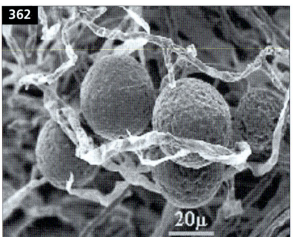

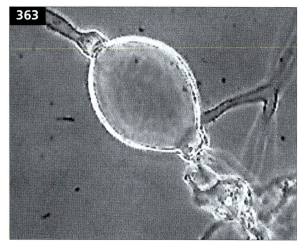

360–363 *Phytophthora cinnamomi*: **360**, **361** *Eucalyptus marginata* forest damaged by the pathogen in Western Australia; **362**, **363**, chlamydospores and sporangium, respectively. (Photos copyright of D. Chavarriaga and S. Woodward.)

ARMILLARIA (HONEY FUNGUS; BOOTLACE OR SHOESTRING FUNGUS)

Armillaria species are probably the most ubiquitous fungal root diseases of woody plants (370–375) and are common in old growth forests, woodlands, arboreta, and orchard crops throughout the world.

Until the 1970s, *Armillaria* was considered a single, variable species (*A. mellea*), but over 40 different species are now recognized. A few of these are very virulent; other species may attack a tree that is seriously weakened by old age or other factors, such as repeated or prolonged water-logging or serious insect infestations in the crown. Knowledge of the effects of the different species is poor and most accounts are generalized.

364–369 Other *Phytophthora* species important on forest trees: **364**, *Castanea sativa* attacked by *P. cambivora*; **365**, dead riverside *Alnus glutinosa* killed by *Phytophthora alni*; **366**, dying group of *Chamaecyparis lawsoniana* in Oregon, following attack by *P. lateralis*; **367**, bark tissues of *C. lawsoniana* showing lesion caused by *P. lateralis*; **368**, death of *Lithocarpus densiflorus* (tan bark oak) caused by *P. ramorum* infection; **369**, dieback of *Rhododendron* infected with *P. ramorum*. (Photos **366–369** copyright of E. Goheen, USDA Forest Service.)

370–375 Symptoms of *Armillaria* infection of secondary roots:
370, mycelial sheath of *A. ostoyae* in the vascular cambium
region of a dead *Picea sitchensis*; **371**, resinosis on the stem of an
infected *Picea sitchensis*; **372**, sub-cortical rhizomorphs (photo copyright of S. Murray); **373**, large clusters of fruiting
bodies at the base of a dead *Picea abies* (photo copyright of W. Bodles). A small *Fomitopsis pinicola* bracket can be seen on
the stem; **374**, fruiting bodies of *A. luteobubalina* at the base of an infected *Eucalyptus diversicolor* (photo copyright of
R. Robinson); **375**, a stand of *Pinus sylvestris* showing an infection centre with dead trees.

376–378 Symptoms of *Heterobasidion* infection of secondary roots: **376**, fruiting body of *H. annosum* at the base of a severely infected *Picea sitchensis*; **377**, lifting pine stumps with a tracked vehicle in East Anglia, England; **378**, decay symptoms in *Picea abies* butt log.

Symptoms

As with other root diseases, a general discoloration and deterioration of the whole crown may be observed. A white mycelial sheath develops under the bark in the collar region, sometimes reaching 1 m up the trunk (**370**). In conifers, particularly *Picea* spp. (spruce), copious resin exudation (resinosis) occurs from the trunk (**371**). If butt rot results, the decayed wood is characteristically wet and stringy.

The common names of 'bootlace' or 'shoestring' fungus relate to the formation of reddish-brown or black cords known as rhizomorphs, by most temperate *Armillaria* species (**372**). Rhizomorphs radiate from infected roots into the surrounding soil and, together with root contacts, are the mechanisms of spread for most temperate *Armillaria* species.

Some African and Australasian species do not produce rhizomorphs in nature, spreading by root contacts alone. Rhizomorphs are extremely resilient and, as long as they remain attached to a substantial food base, can withstand adverse environmental conditions. Toadstools of *Armillaria* (**373, 374**) are of limited use in diagnosis because of their transient nature and the similarity between species.

The size of certain *Armillaria* infections has been investigated in detail in a few locations, particularly in North America. The largest single genotype recorded to date is an *A. ostoyae* in Oregon. Here, it occupies over 965 ha of coniferous old growth forest, making it the largest living organism known, with a probable age of over 2,000 years.

Table 5 Some important diseases of secondary roots of trees

Species	Locality	Principal hosts	Damage	Figures	Comments
Armillaria	Ubiquitous	Wide host range	Root decay; cambial death; tree death; wet rot	370–375	Many different species, some saprotrophic, others virulent pathogens
Heterobasidion	North Temperate gymnosperm forests	Wide host range	Root decay, butt rot; kills pines	376 378, 404	Three species in Europe; 2 in North America; other saprotrophic species in Asia
Phellinus weirii	Pacific north-west, Russia, Japan	*Pseudotsuga menziesii, Tsuga mertensiana, Abies grandis*; other Pinaceae	Root decay, reduced leader growth; laminated root rot; tree death	379, 380	Probably two distinct species of *Phellinus* involved in North America and Asia
Phellinus noxius	Tropical	Wide host range in angiosperms and gymnosperms	Root decay; resinosis; host death	383	Dark brown fungal crust on root and butt surfaces; causing major problems on *Cordia alliodora* in Vanuatu
Rigidoporus lignosus	Tropical	Angiosperm trees	Root decay	–	White root rot
Inonotus tomentosus	Boreal forests of northern hemisphere	Pinaceae, particularly *Picea* and *Pinus*	Reduced increment due to root loss; host death; windthrow	381	Serious disease in central interior forests of British Colombia, Ontario and Quebec
Phaeolus schweinitzii	North temperate gymnosperm forests; South Pacific	Pinaceae; occasional on angiosperm trees	Extensive brown cubical rot; tree collapse	412	Root infection may be predisposed by prior root killing by *Armillaria* spp.
Rhizina undulata	Wide distribution	Wide host range in gymnosperms	Kills cambium; host death	382	Dependent on fires for spore germination; does not cause decay

Economic importance

Major losses are associated with replanting on old forest sites, particularly where the plantation species is different to the naturally occurring dominant tree type. Young conifers, 20–30 years old, may be particularly susceptible to attack on former hardwood sites. More severe disease can occur if site conditions are unfavourable. Damage in arboreta, parks, and private gardens may cause concern. Chronic infections persist, killing a proportion of the root system and reducing growth increment.

379–383 Symptoms of other fungal diseases causing death of secondary roots: **379**, laminated rot in *Pseudotsuga menziesii* root wood caused by *Phellinus weirii*; **380**, disease transfer of *P. weirii* has occurred, via root-to-root contact, between a dead *P. menziesii* (right) and an adjacent sapling (photo copyright of E. Goheen, USDA Forest Service); **381**, fruiting body of *Inonotus tomentosus* arising from a subterranean root of *Picea abies* (photo copyright of W. Bodles); **382**, fruiting bodies of the discomycete root pathogen *Rhizina undulata*; **383**, crust formed by *Phellinus noxius* on the lower trunk of *Delonix regia* (photo copyright of C. Hodges).

Management

As *Armillaria* grows within woody tissues, often below ground, control is difficult. Chemicals, even those marketed for control of *Armillaria*, are ineffective. Management measures applied with variable success include physical removal of stumps by winching, chipping, or trenching around infected trees and letting an impermeable barrier, such as heavy gauge plastic sheet, into the ground to prevent spread of rhizomorphs.

In Californian citrus orchards, soil sterilization is used to control *Armillaria*. Old stumps are removed before applying the treatment. Although the chemicals used do not kill all of the *Armillaria* in the remnants of the root systems, other fungi, particularly *Trichoderma*, colonize the soil effectively in the absence of competitors, and *Armillaria* is unable to grow out of woody debris. Potassium phosphite injection has shown effectiveness against *Armillaria* in orchard crops in Australia.

Anecdotal evidence has suggested that different tree species vary in susceptibility to *Armillaria*. However, much of the evidence requires re-evaluation in the light of new information on speciation in the genus.

HETEROBASIDION ANNOSUM (FOMES ROOT AND BUTT ROT, ANNOSUS ROOT ROT)

Economically, species of *Heterobasidion* are the most important pathogens of managed gymnosperm forests in the north temperate zone, causing death of trees and serious decay. The disease has also been serious in the southeast of North America. Three species of *Heterobasidion* are found in Europe, and two putative species are recognized in North America. European species tend to occur on particular host genera: *Pinus* (pine), *Picea* (spruce) or *Abies* (fir), but are not completely restricted to these hosts. The pine group, for example, has a very wide host range, including other coniferous genera and angiosperm trees. Different species of *Heterobasidion* occur elsewhere in the world, but are generally regarded as saprotrophic, causing decay of trees killed by other agents.

Symptoms

Symptoms vary with the tree species concerned. An obvious sign of attack is the presence of the hard, leathery perennial fruit body at the base of a dying or dead tree (**376**), or on the roots of wind-blown trees. The bracket is up to 30 cm across, with a reddish-brown upper surface, becoming dark-brown with age, and a white–cream poroid underside. Fallen conifer needles and twigs may be surrounded by the growing fruit body. Spores are released whenever the temperature is above freezing, remain viable for long periods, and are fundamental in initiating infections.

Extensive butt rot develops in susceptible species, initially as a red stain visible in the stem when the tree is felled. Later the wood becomes pale with black flecks, followed by small lens-shaped pockets of white material (incipient decay). The white pockets of fungal material (visible in **378, 404**) are a diagnostic feature of *Heterobasidion* decay, and eventually they coalesce. Advanced decay is dry and stringy. By the time the trees are 30 years old, decay may extend 4 m up the stem. Affected trees are liable to snap in high winds.

In pines butt rot is rare except in old age. Symptoms of infection are shortening of shoots and needles, reduction in foliage density, and general discolouring of the crown. Tree death is common on free-draining, alkaline sandy soils in low rainfall areas, but is less likely on more acidic mineral soils.

Disease cycle

Initial infection occurs when spores are deposited on a freshly exposed stump surface. Following germination, the fungus penetrates into the dying root system, where further spread occurs at root contacts between stumps and roots of adjacent trees, and subsequently between standing trees.

New infections arise in each plantation thinning, with the exposure of fresh stumps increasing disease incidence and severity during the first rotation. If, after felling, the site is replanted with gymnosperms, the second crop becomes infected when the roots contact colonized woody material in the soil. *Heterobasidion* persists in woody material in the soil for many years, as illustrated by its recovery from larch stumps 63 years after the trees were felled. This reservoir of infection provides continuity in disease between generations of trees on affected sites.

Economic importance

Losses result from reduction in increment and useable timber, and from wind-blow of trees exposed due to killing or snapping of infected adjacent individuals. Hosts other than pine may also be killed on sites with soil pH above 6, or in areas with very high infection rates in previous rotations.

In 1998, it was estimated that *Heterobasidion* caused annual losses to forestry in the European Union of approximately €700 million. Some other European countries also have a high incidence of infection.

Management

The disease is particularly amenable to control at the time of the initial spore infection on stump surfaces. Management methods available include chemicals, biological agents, and physical techniques.

Stumps may be treated with 20% aqueous urea solution (30% during mechanized harvesting) immediately after trees are felled. Ammonia gas, liberated as urea degrades, increases the stump surface pH to levels at which *Heterobasidion* spores cannot germinate. Once all the urea has degraded, the stump surface is usually too dry for *Heterobasidion* spores to germinate. Moreover, other competitive non-pathogenic fungi colonize the stump surface in that time and out-compete any further *Heterobasidion* spores that may alight on the stump. Other chemicals used in stump treatment have included creosote, sodium nitrite, ammonium sulphamate, paraquat, and di-sodium octaborate. In North America, borax is dusted onto the stump surface to prevent basidiospore infection.

Biological control, using spores of the saprotrophic decay fungus *Phlebiopsis gigantea* painted onto the stump surface, has also been employed successfully. *P. gigantea* grows very rapidly into the stump and out-competes any *Heterobasidion*. Originally developed for use on pine in the east of England, further strains of the fungus are now produced commercially for use on spruce.

On flat sites with light soils and a pH greater than 6, it may be economically viable to physically remove stumps. A tine on a hydraulic arm attached to a vehicle is used to lift stumps out of the soil (377).

The stumps are then bulldozed into stacks at 8 m intervals and left to dry and decay. Under these conditions, *Heterobasidion* dies out rapidly, and this technique significantly reduces death and decay in subsequent rotations.

Restricting felling to periods when the temperature is above 27°C prevents infection because the fungal spores die. This method can be used in the southeastern states of the USA during May to September, but places restrictions on forest operations, creating problems in management and marketing. Delayed replanting, or replanting sites with broadleaved trees, is used in some localities to reduce the impact of *Heterobasidion*.

DISEASES OF STEMS

CANKERS

Cankers (384–393) are sunken lesions on stems or branches, which form as a result of pathogens infecting bark tissues, usually through natural cracks, leaf scars, or wounds (see also Chapters 6, 12). Two basic types of tree canker are defined, namely perennial (regular) and diffuse. Examples of both are listed in *Table 6*.

Development of perennial cankers follows growth of the host. During the growing season, the host develops barriers around the canker, producing concentric rings of callus. This cycle of growth may be repeated over many years, resulting in the formation of typical 'target cankers', such as those caused by *Lachnellula willkommii* on *Larix decidua* (386) or by *Nectria galligena* on a range of angiosperm trees (390).

Diffuse cankers grow more rapidly than perennial cankers, as the causal agent is able to overcome or inhibit formation of host barriers during the growing season. With rapid growth rates, some cankers may girdle affected twigs and branches within a growing season. Cankers typical of this type are *Seiridium* canker of *Cupressus* species (387) and *Cryphonectria* canker of *Castanea* (388).

Lachnellula willkommii (larch canker)

Canker is a major disease of *Larix decidua* in Europe and is also found in Japan and North America.

Infection biology

Canker formation is typical of perennial cankers (386). Characteristic small circular orange fruiting bodies of the pathogen are produced on the surface of the cankers which develop on the branches and main stem. Recovery from canker occurs when the fungus fails to breach the periderm formed in the growing season.

Losses

Many *Larix* spp. (larches) are killed by *L. willkommii*. In a tree that survives, serious internal deformation can render the whole tree useless for timber production. Although seriously affecting only *L. decidua* in Europe, canker has been recorded on *L. kaempferi* and *L. sukaczevii*. In general, *L. kaempferi* and hybrid larch are quite resistant to *L. willkommii*.

Management

Less-susceptible provenances, resistant species, and hybrids may be planted; Sudetenland provenances of *L. decidua* show greater resistance to canker than alpine and lowland provenances. In North America, attempts are being made to eradicate larch canker.

Seiridium cardinale (coryneum canker of cypress)

Cypress canker is the most important disease of *Cupressus*, causing problems wherever these trees are grown for timber or for amenity. The sexual state of the pathogen (*Leptosphaera*) is known only from California.

Infection biology

The pathogen enters through small wounds and grows slowly through the bark, killing tissues; resin bleeding (resinosis) occurs from infected bark. Branches become girdled, with eventual death of distal parts (387), and the foliage fades to yellow and finally to brown before falling. Continued growth of the tissue around the lesion leads to the formation of long, sunken cankers containing pin-head sized fruiting bodies (acervuli). The spores spread in water droplets and rain, but probably only over short distances.

384–393 Canker diseases of trees: **384**, dieback in *Pinus nigra* plantation, northern Alps, caused by *Sphaeropsis sapinea*; **385**, dieback of young *P. nigra* caused by *Gremmeniella abietina* attack (photo copyright of S. Murray); **386**, canker caused by *Lachnellula willkommii* on *Larix decidua*; **387**, dieback of *Cupressus sempervirens* caused by *Seiridium cardinale*; **388**, canker on *Castanea americana* caused by *Cryphonectria parasitica* (photo copyright of G. Griffin); **389**, basal canker on *Eucalyptus tereticornis* possibly caused by *Cryphonectria cubensis* (photo copyright of E. Boa); **390**, *Nectria galligena* regular canker on cultivated apple; **391**, leaf symptoms of *Apiognomonia veneta* on *Platanus* x *acerifolia*; **392**, crown of *Salix alba* infected with *Venturia saliciperda*; **393**, diffuse canker on *Fraxinus excelsior* caused by *Pseudomonas syringae* s.sp. *savastanoi*, pv. *fraxini*.

Losses

The disease causes serious dieback on *C. sempervirens* in Mediterranean countries where cypress is a very important landscape tree. In East Africa, *C. macrocarpa* and *C. sempervirens* were discarded for timber production following a serious outbreak of *Seiridium* canker.

Variation in susceptibility occurs within and between species in the family Cupressaceae. *C. macrocarpa* and *C. sempervirens* are particularly susceptible to *Seiridium*. *Chamaecyparis lawsoniana*, *C. nootkatensis*, *Cupressus arizonica*, *C. glabra*, and *C. lusitanica* are very resistant.

Table 6 Examples of important canker and other shoot diseases of trees

Species	Locality	Principal hosts	Damage	Figures	Comments
Sphaeropsis sapinea	Worldwide	*Pinus* spp.	Death of current year's shoots	384	Common, but aggressive in warmer climates; can be severe in 10–40-year-old plantations; occasional on seedlings
Gremmeniella abietina	Northern hemisphere	Pinaceae	Necrosis of bud base; death of shoots; premature needle loss; bark damage	385	Very destructive disease; development associated with poor choice of species for site and climate, but syndrome possibly more complex
Lachnellula willkommii	Europe; rare in North America and Japan	*Larix* spp., esp. *L. decidua*	Cankers on branches and stems; girdling and dieback; orange disc-shaped fruiting bodies, 2–6 mm diameter	386	Significant disease of European larch; most troublesome in lowland and alpine provenances
Seiridium cardinale	World-wide	Cupressaceae	Dieback, host death	387	Most serious disease of *Cupressus* spp.
Cryphonectria parasitica	North America, southern Europe	*Castanea* spp.	Stem cankers; dieback	388	Chestnut blight; wiped out industries based on *Castanea* in North America
Cryphonectria cubensis	Tropical and sub-tropical regions world-wide	*Eucalyptus* spp., particularly *E. saligna*, *E. grandis*, *E. tereticornis*	Wound pathogen; large cankers at base of stem; stem may be girdled	389	Serious threat to tropical plantations of *Eucalyptus*
Hypoxylon mammatum	North America; sporadic in Europe	*Populus* spp., particularly *P. tremuloides*, *P. tremula* and their hybrids	Cankers on bark of stems; dieback of branches; premature death	–	Most serious disease of poplars in North America
Nectria galligena	Europe, North America, Chile	Angiosperm trees, including species of *Fagus, Fraxinus, Malus, Pyrus, Acer, Populus*	Regular cankers on branches; dieback; red perithecia on canker margins	390	Very widespread pathogen; particular problem in apple and pear orchards
Apiognomonia veneta	Europe, North America	*Platanus* spp., especially *P. occidentalis* and *P. x acerifolia*	Dieback of twigs; leaf lesions	391	Occasional severe attacks associated with moist spring conditions. *P. orientalis* fairly resistant

Species	Locality	Principal hosts	Damage	Figures	Comments
'Beech Bark Disease'	Europe, North America	*Fagus* spp.	Tarry spots on bark; dimpling of bark; death of vascular cambium; large necrotic areas on bark	–	Disease complex caused by infestation by felted beech coccus, *Cryptococcus fagisuga*, followed by *Nectria coccinea* infection. Common in young plantations
Cryptostroma corticale	Northern Europe	*Acer pseudoplatanus*	Green stain in sapwood; wilting; dieback; black sporulation beneath dead bark	–	Outbreaks appear in years following warm, dry summers
Venturia saliciperda	Europe, North America	*Salix fragilis*, *S. alba*, *S. viminalis*	Necrotic black blotches on leaves progressing onto shoots; withering of foliage; dieback	392	Willow scab; symptoms easily confused with those of black canker (*Glomerella miyabeana*)
Cryptodiaporthe populnea	Europe, eastern North America	*Populus* spp., particularly *P. nigra*	Bark necroses at bases of side branches; dieback	–	Frequent on *P. nigra Italica*
Pseudomonas mors-prunorum	Europe, Africa	*Prunus* spp.	Shot-holes in leaves; depressed lesions on branches and stems with gum-like ooze; wilt; dieback	–	Bacterial canker of cherry
Pseudomonas syringae sub sp. *savastanoi* pv. *fraxini*	Europe	*Fraxinus*; other Oleaceae	Rough, irregular cankers, variable in size, on bark of branches and stems	393	Bacterial canker; most damaging disease of ash in Europe. Similar to pathogen causing olive knot
Erwinia amylovora	Europe, North America, New Zealand	*Malus* spp., *Pyrus* spp., *Sorbus* spp.	Rapid blackening and wilting of flowers and leaves; death of very susceptible species	–	Fireblight of Rosaceae, sub-family Pomoideae. Bacterial disease. Variation in susceptibility exists between host varieties and species
Xanthomonas populi	Northern Europe	*Populus* spp. and hybrids	Irregular cankers on branches and stems; dieback	–	Bacterial canker of poplar. Great variations in susceptibility are found between poplar clones

Management

The application of fungicides is impractical. Pruning out infected parts prolongs the life of the tree, but in badly affected areas the replacement of infected trees with resistant species is the only effective treatment. Selection and breeding programmes against *S. cardinale* are under way in Italy and Greece.

CRYPHONECTRIA PARASITICA (CHESTNUT BLIGHT)

Following its introduction into North America around 1900, on infected planting stock of Asian *Castanea* species, chestnut blight has been responsible for the demise of the American chestnut (*Castanea dentata*) for timber production. In the late 1930s the disease was inadvertently introduced into Italy and Yugoslavia, affecting the European sweet chestnut, *C. sativa*, and is spreading northwards into southern France and Switzerland.

Infection biology

The spores are dispersed by rain splash and wind. Lesions, with abundant small pale-brown fruiting bodies, develop on the branches and main stems (388). These eventually become girdled and die back.

Management

A decline in severity of the disease occurred in Italy and Yugoslavia when a mycovirus ('d' factor) infected the pathogen. Cankers caused by mycovirus-infected strains of *C. parasitica* are less damaging.

The importation of *Castanea* planting stock and barked timber into unaffected areas is now prohibited.

NECTRIA GALLIGENA (*NECTRIA* CANKER)

Nectria galligena causes canker of apple and pear in orchards and also affects other angiosperm trees. Regular, target cankers are formed on the branches and main stems of affected trees (390).

Infection biology

The pathogen may enter bark through pruning wounds, and swollen bark develops around infection sites. Small red perithecia of *Nectria* develop in clusters on the surface of these cankers.

Losses and management

In northern Europe, beech and ash are commonly affected by this fungus but it is rarely necessary to control the disease outside the orchard. Cankered branches can be pruned off the tree.

PHYTOPHTHORA RAMORUM: SUDDEN OAK DEATH

In the last 15 years, there has been increasing concern over the potential for *Phytophthora* species to cause serious damage to forest ecosystems, as several newly discovered species of *Phytophthora* have been reported. Particularly noteworthy is the destruction caused to trees of the oak family in California and Oregon due to infection by *P. ramorum* (368, 369; *Table 4*). This pathogen is causing great concern, and is the subject of intensive quarantine measures throughout the world. It is present in Europe, where the pathogen appears to attack mainly hardy ornamental nursery stock, such as species of *Rhododendron* and *Viburnum*, although in the early years of the 21st century, the pathogen has also been found on diseased species of *Quercus* (oak), *Fagus* (beech), and *Aesculus hippocastanum* (horse chestnut) in The Netherlands and the UK. On ornamental shrubs, the disease is often restricted to leaf blotching and dieback of affected shoots, whereas on many tree species, death of the crown occurs following the development of cankering-like symptoms on the secondary tissues of the stem or branches. In California, the problem was first noted in the mid-1990s, when oak family trees began dying in natural forests in coastal counties. *Lithocarpus densiflorus* (tan bark oak) and related species were particularly badly affected. Since these initial reports, *P. ramorum* has been found causing disease on a wide range of host woody plants, including species of *Quercus*, *Umbellaria*, *Pieris*, *Vaccinium*, and *Camellia*, and on *Sequoia sempervirens* (coastal redwood). The potential for *P. ramorum* to cause serious damage to tree and shrub species on continents outside North America has yet to be evaluated in full. The implications for disease management, if *P. ramorum* did become more widespread, could prove enormous.

BACTERIAL CANKER OF POPLAR

Bacterial canker is the most important disease affecting *Populus* spp. (poplar) in Europe, but it is absent in North America. The causal agent, *Xanthomonas populi* s.sp. *populi*, is host specific, attacking only susceptible poplar species.

Disease cycle and symptoms

The bacterium is probably transmitted by rain splash, on the prevailing wind, and by insects. Entry to host trees is probably via wounds, but bacteria may also gain entry via nectaries, hydathodes, and lenticels. It grows in the vascular cambium and a whitish slime fills the intercellular spaces and exudes from small cracks on infected branches in spring. Affected branches may be girdled and die back but on older trees cankers can be present for many years. Here, large and diffuse cankers develop due to the formation of callus around the lesion margin.

Losses caused

Infection may lead to severe dieback; even small numbers of cankers on a trunk render it useless for timber.

Management

The most effective way of managing canker is to plant resistant clones. *Populus nigra* and many *P.* x *euamericana* hybrids are completely resistant, whereas some clones, e.g. 'Brabantica' and 'Grandis', are extremely susceptible to bacterial canker. The temperature-sensitive nature of the bacterium prevents serious disease development in warmer parts of Europe. Legislation in some European countries requires the removal of diseased trees as the source of inoculum.

DISEASES OF FOLIAGE

Although ageing and senescent foliage of both gymnosperm and angiosperm trees may be attacked by opportunistic pathogens, the critical damage occurs when young foliage is seriously attacked during rapid growth periods (*Table 7*, **394–402**).

Many diseases of foliage are characterized by spotting. These spots may be restricted in size or may expand rapidly until large areas of the leaf surface are damaged. Sometimes spots are bordered by discrete, dark zone lines but they may also expand unevenly over the needle or leaf. Some examples of foliage diseases are described below, but many other foliage diseases occur (*Table 7*).

RED BAND NEEDLE BLIGHT OF PINE (*MYCOSPHAERELLA PINI*)

Red band needle blight (Dothistroma blight) attacks may species of pine, but is a particular problem on *Pinus radiata* (Monterey pine, **394, 395**) and *P. ponderosa* (western yellow pine). It caused very serious damage in East Africa, New Zealand, and Australia in the 1950s. In East Africa, *P. radiata* is no longer grown on a commercial scale because of the disease.

Symptoms

Dark red banding appears on the needles, leading to their early abscission and consequent loss in tree growth increment. Repeated severe attacks can lead to tree death.

Management

This disease provides one of the few examples of the economically viable use of fungicides in controlling a forest pathogen. Red band needle blight is treated by aerial spraying with fungicides in Australia and New Zealand.

NEEDLE CASTS OF CONIFERS

Defoliation of conifers caused by needle cast diseases results in growth reduction and, in extreme cases, death. Many fungi are associated with defoliation, including: *Lophodermium seditiosum*, *L. sulcigena*, *Mycosphaerella* (*Dothistroma*) *pini*, and *Cyclaneusma minus* on *Pinus* spp. (pine); *Lophodermium piceae* and *Rhizosphaera kalkhoffii* on *Picea* spp. (spruce); *Rhabdocline pseudotsugae* on *Pseudotsuga menziesii* (Douglas fir); and *Mycosphaerella laricina* and *Meria laricis* on *Larix* spp. (larch). *Phaeocryptopus gauemannii* causes Swiss needle cast of Douglas fir, considered a particular problem in Christmas tree plantings, but more recently raising concerns in Douglas fir-dominated forests in the Pacific North West.

LOPHODERMIUM NEEDLE CAST OF PINES

A range of *Lophodermium* species occurs on foliage of two-needle pines. *L. seditiosum* causes serious needle cast of *Pinus sylvestris* (Scots pine) in nurseries (**396**) and in plantations. It is also common on older trees, where damage appears insignificant, although effects on growth have not been quantified. However, older trees act as disease reservoirs for nearby nurseries. The symptoms of *L. seditiosum* are easily confused with those of two other *Lophodermium* species of little importance on living needles. *L. pinastri* occurs on abscised needles, whereas *L. conigenum* proliferates on needles attached to fallen branches.

Losses

The heaviest losses of pine seedlings occur in nurseries adjacent to pine plantations, which are the main sources of fungal inoculum.

Management

L. seditiosum can be controlled in nurseries using fungicide sprays. As the spores are viable only over short distances, infection can be avoided completely if nurseries are sited 2–3 km away from mature pines.

POWDERY MILDEWS

Mildews do not often have a wide host range but are usually host specific, attacking a limited number of species within a genus. The symptoms result from the white or pale-grey growth (largely superficial and often with a powdery texture) forming on the surfaces of leaves and young shoots. The pathogen withdraws nutrients from the host via haustoria which develop inside the host cell wall, invaginating the host cell plasmalemma.

A significant powdery mildew in European forests is that caused by *Microsphaera alphitoides* on *Quercus* spp. (oak, **397**, **398**) and, rarely, *Fagus sylvatica* (beech) and *Castanea sativa* (sweet chestnut). *Uncinula* (*Sawadea*) *bicornis* and *U. tulasnei* attack *Acer pseudoplatanus* (sycamore, maple) and *A. platanoides* (Norway maple), respectively (**399, 400**), sometimes causing significant damage in amenity situations.

Management

Where necessary, powdery mildews are controlled using fungicide sprays.

TAPHRINA DEFORMANS (PEACH LEAF CURL)

Leaf curl (**402**) is a common problem on *Prunus persica* (peach), and closely related species, in most areas where these trees are grown. *Taphrina deformans* infects leaves in early spring, causing growth abnormalities. These 'blisters' usually turn pale pink to bright red and the affected leaves abscise in early summer. The pathogen exists as a free-living organism on leaves and shoots in the summer months, and as spores during winter.

Taphrina is controlled in orchards with fungicides, spraying just before bud burst and at 14-day intervals thereafter. The infected leaves should be gathered and burned.

Table 7 Examples of foliage diseases of trees

Species	Locality	Principal hosts	Damage	Figures	Comments
Phaeocryptopus gauemannii	North America, Europe, Australasia	*Pseudotsuga* spp.	Pale-yellow foliage; premature defoliation beginning with older needles; reduced increment	–	Particular problem where host planted in unsuitable conditions. *Rhabdocline pseudotsugae agg.* cause similar gross symptoms
Mycosphaerella pini	World-wide	*Pinus* spp.	Pale brown bands around needles, with dark red-brown margins; premature defoliation; reduced increment; death	394, 395	Red band needle blight. Very serious disease in plant-ations of *Pinus radiata* in East Africa, Chile and New Zealand
Lophodermium seditiosum	Europe, North America	Two- and three-needle *Pinus* spp.	Death of affected foliage; elongated black fruiting bodies (apothecia) on needles	396	Particular problem in nurseries and on young planting stock; older trees also attacked
Microsphaera alphitoides	Europe	*Quercus* spp.	White powdery patches on upper leaf surface; affected leaves distorted; sexual stage (cleistothecia) form as tiny orange dots, turning black, in late summer	397, 398	Powdery mildew. Affects plants of all ages
Uncinula tulasnei	Europe	*Acer platanoides*	White powdery spots on upper leaf surface may coalesce to cover whole leaf; in late summer cleistothecia form as tiny orange dots, turning black	399, 400	Powdery mildew. *Uncinula bicornis* attacks *Acer pseudoplatanus* and *A. campestre*; affects both sides of leaves
Pleuroceras pseudoplatani	Europe	*Acer pseudoplatanus*	Large brown lesions on leaf, often around point of attachment of petiole to lamina	401	Giant leaf blotch of sycamore maple

(Continued)

Table 7 Examples of foliage diseases of trees (*Continued*)

Species	Locality	Principal hosts	Damage	Figures	Comments
Rhytisma acerina	Europe, North America	*Acer* spp., particularly *A. pseudoplatanus*	Black shiny lesions, 1–2 cm diameter on leaves	–	Tar spot. Very common disease wherever susceptible *Acer* spp. grow
Drepanopeziza punctiformis	Europe, North America	*Populus nigra, P. deltoides,* hybrids between these species	Dense dark brown spotting on leaf, coalescing. Premature defoliation. Reduced growth, dieback	–	Marssonina leaf spot. Weakens badly affected trees which then are more susceptible to secondary pathogens
Taphrina deformans	Europe, North America, Australasia	*Prunus persica, P. amygdalus* and related spp.	Distortion of leaf lamina, red blisters, defoliation in late spring	402	Peach leaf curl. Very common in peach and almond growing areas

394–402 Diseases of foliage: **394**, severe *Mycosphaerella pini* infection in the lower crown of *Pinus radiata* in New Zealand; **395**, close-up of *P. radiata* seedling showing red banding on the needles due to *M. pini* infection; **396**, severe needle browning on *P. sylvestris* caused by *Lophodermium seditiosum* (photo copyright of S. Murray); **397**, leaf symptoms on *Quercus robur* (oak) caused by the powdery mildew *Microsphaera alphitoides*; **398**, cleistothecia of *M. alphitoides* with regularly branched appendages; **399**, leaf symptoms of *Uncinula tulasnei* (maple mildew) on *Acer platanoides* (Norway maple); **400**, cleistothecia of *U. tulasnei*, the appendages of which lack branches; **401**, large leaf lesions caused by *Pleuroceras pseudoplatani* on *Acer pseudoplatanus*; **402**, typical red blisters and growth distortions on leaves of *Prunus amygdalus* infected with *Taphrina deformans*.

DECAY

Decay of the roots and stems of trees causes the greatest economic losses in forestry. Apart from direct timber losses, it causes tree instability and leads to potential safety problems. Decay should, however, be considered in its true natural and essential ecological role: it releases nutrients that have been locked up in lignified tissues for many years. This process removes woody debris from the forest ecosystem, increasing soil fertility and organic matter content. Decay holes and decaying wood also provide important habitats for other organisms.

Much decay results from the growth of fungi which are restricted to the heartwood during the lifetime of the tree, while few fungi grow within functional sapwood. Many entirely saprotrophic fungi degrade woody debris on the forest floor, and others attack processed timber.

Decay is caused by highly specialized fungi, usually in the order Hymenomycetes, although a few Ascomycotina in the Xylariaceae also degrade lignin. Some selected decay-causing fungi are listed in *Table 8*, and are illustrated in **403–412**.

Decay can be divided into brown and white rot, according to the type of decomposition. Brown rot fungi produce cellulases, polygalacturonases (pectinases), and xylanases, degrading carbohydrate polymers in cell walls; lignin is not decomposed. In advanced brown rots, the wood has a dry and

Table 8 Examples of important decay-causing fungi of the order hymenomycetes, and their niches. Several other decay-causing species are included in Table 5

Species	Locality	Principal hosts	Damage	Figures	Comments
Stereum spp.	Ubiquitous	Gymnosperm and angiosperm trees	Wound decay; white rot, cankers; small bracket-like or resupinate fruit bodies lack pores`	–	Range of species, including those in *Amylostereum* and *Haematostereum*
Chondrostereum purpureum	Worldwide	Wide host range, mostly angiosperm trees; occasional on gymnosperms	White rot; dieback; host death	–	Aggressive decay pathogen, cause of silver leaf disease; particular problem on trees in the Rosaceae
Ganoderma spp.	Ubiquitous	Very wide host range in angiosperms, including Palmae	White rot in angiosperm trees; root rot of palms in plantations	405	Taxonomy is confused
Fomes fomentarius	Europe, North America, Asia	Angiosperm trees, particularly *Fagus, Betula*	White rot; characteristic layering of hoof-shaped fruit body, with dark grey upper surface	406	Tinder fungus; extensive decay of heartwood
Polyporus squamosus	Europe, Asia, Australia, rare in North America	*Ulmus, Acer, Juglans*	White rot; large fan-shaped fruit body with pale fawn upper surface	407	Common in Europe

Species	Locality	Principal hosts	Damage	Figures	Comments
Bjerkandera adusta	Europe, North America	Angiosperm trees; also colonizes stumps of gymnosperms	Wound decay	–	Common on *Fagus* spp.
Meripilus giganteus	Europe, North America	*Fagus* spp.; rarely other angiosperm trees	Decay of upper root system; large pale–dark brown fleshy, annual fruiting bodies at base of tree	408	Decay of root plate leads to instability
Phellinus pini	Northern hemisphere gymnosperm forests	Wide host range on gymnosperm trees	White rot; perennial bracket-like fruit body dark brown on upper surface	–	Considered one of the most serious heart rot fungi in North America
Pleurotus ostreatus	Europe, North America, Africa, Asia, Australasia	Angiosperm trees, notably *Fagus, Quercus, Aesculus, Populus*	White rot is flaky; annual gilled fruiting bodies	–	Oyster fungus; extensive decay may occur
Laetiporus sulphureus	Europe, North America	Wide host range, including both angiosperm and gymnosperm trees	Brown cubical rot; large yellow, annual fleshy fruiting bodies	409	Common on *Quercus* spp.; also found on *Taxus*
Fomitopsis pinicola	Europe, North America	Angiosperm and gymnosperm trees	Brown cubical rot; bracket-like fruiting bodies are red-brown on upper surface, with a yellow–pale brown pore layer	410	Significant in old-growth forests of western North America; common in forests of continental Europe
Sparassis crispa	Europe, North America, Japan	Pinaceae	Brown cubical rot; large, fleshy annual fruit bodies, cauliflower shaped	411	Decayed stems may snap in high winds
Phaeolus schweinitzii	Europe, North America, Central America, Japan, East Asia, South Africa, New Zealand	Pinaceae; occasional on angiosperm trees	Brown cubical rot; large annual fruiting bodies produced on stem or near stem base	412	Decayed stems may snap in high winds; in North America trees are killed by infections

162

403–412 Decay and fruiting bodies of decay-causing fungi:
403, brown rot in *Picea abies* – note the regular cubical
cracking in the decayed wood (photo copyright of W.
Bodles); **404**, severe white rot in stem of *Picea abies* decayed
by *Heterobasidion annosum*; **405**, *Ganoderma applanatum*
fruiting bodies (photo copyright of W. Mulenko, Marie-
Curie-Sklodowska University, Lublin, Poland); **406**, *Fomes
fomentarius* on the stem of *Betula pubescens*; **407**,
Polyporus squamosus on *Acer pseudoplatanus* (photo
copyright of S. Thompson); **408**, *Meripilus giganteus* at
base of *F. sylvatica*; **409**, *Laetiporus sulphureus* on a fallen
log of *Quercus robur*; **410**, *Fomitopsis pinicola* on the stem
of *Picea abies* (photo copyright of W. Mulenko); **411**,
fruiting body of *Sparassis crispa* at the base of *Pinus
sylvestris*; **412**, fruiting body of *Phaeolus schweinitzii* on
wind-snapped *Pseudotsuga menziesii*.

crumbly appearance, often with marked cracking,
which may be of a cubical nature (**403**). Sheets of
mycelium may develop in the cracks.

White rot fungi, in addition to the enzymes
described above, also produce several lignin-
degrading enzymes (laccases, lignin peroxidases,
and manganese-dependent lignin peroxidases). The
peroxidases oxidatively cleave phenolic compounds
from the lignified structure of the cell wall. White
rot species differ in their relative abilities to degrade
lignin and carbohydrates. Decay may be stringy,
flaky, or pocketed (**404**) and is usually dry. In some

fungal species, their ability to degrade early and late wood differs: generally the thinner-walled early wood is more readily degraded and results in a laminated rot (**379**).

MODE OF ENTRY

Classically, decay-causing fungi were thought to enter trees through wounds. Damage to the bark of the tree opens points where spores of decay fungi impact and germinate; the fungus then grows into the heartwood of the tree and, depending on conditions therein, causes rot. A recent controversial hypothesis suggests that some decay-causing species may exist endophytically inside the tree for years without causing any decay. Endophytic fungi can be isolated from sound wood where they appear to exist in stasis, but wounding changes the microenvironment within the tree, allowing air to enter and altering the gaseous balance. Such events then trigger growth of the fungus and result in decay.

Table 9 Examples of important rust diseases of trees

Species	Locality	Principal hosts	Damage	Figures	Comments
Cronartium flaccidum	Europe	2-needle pines, esp. *Pinus sylvestris, P. pinea, P. pinaster*	Dieback of branches; death of host when main stem is girdled	**413**	Alternate hosts: *Vincetoxicum, Paeonia, Tropaeolum.* Microcyclic form: *Endocronartium* (*Peridermium*) *pini*
Cronartium ribicola	Europe; North America	5-needle pines	Dieback of branches; death of host when main stem is girdled	**414–417**	Alternate hosts: *Ribes* spp.
Cronartium f. sp. *fusiforme quercuum*	South-east USA, west to Texas	*Pinus taeda, P. echinata, P. elliottii*	Kills young trees; cankers on branches and stems of older trees lead to breakage	–	Alternate hosts: *Quercus* spp. (red oaks); damage is severe in plantations
Melampsora spp.	Worldwide, where *Salix* spp. are grown	*Salix* spp.	Premature defoliation; reduced increment, dieback	**418, 419**	Several species recognized. Alternate hosts: *Abies, Allium, Euonymus, Larix, Ribes.* Affected trees often infected by secondary pathogens
Melampsora spp.	Worldwide, where *Populus* spp. are grown	*Populus* spp.	Premature defoliation; reduced increment, dieback	**420**	Several species recognized. Alternate hosts: *Allium, Arum, Mercurialis, Pinus, Larix.* Affected trees often infected by secondary pathogens
Melampsoridium spp.	Europe, Asia, Japan, North America	*Betula, Alnus*	Yellow spotting on leaves; reduced increment; severe attacks lead to premature defoliation	**421**	*M. betulinum* attacks *Betula* spp.; *M. hiratsukanum* attacks *Alnus* spp. Alternate host: *Larix* spp.

413–416 Rust diseases of gymnosperms: 413, top dieback of *Pinus sylvestris* resulting from girdling by *Endocronartium pini*, the microcyclic form of *Cronartium flaccidum*; 414, abundant formation of aecidia of *Cronartium ribicola* on branches of *Pinus strobus*; 415, *Ribes nigrum* with a serious attack by *C. ribicola*; 416, yellow uredosori and red teleutosori of *C. ribicola* on underside of a *Ribes nigrum* leaf.

RUST DISEASES

There are approximately 7,000 rust fungi which are obligate pathogens and require a suitable host plant for growth. Some cause diseases of great economic importance to both angiosperm and gymnosperm trees (*Table 9*, 413–421). Rusts have complex life cycles, with up to five different spore types (macrocyclic species, for example *Cronartium ribicola*, 417), usually produced on two unrelated plants. In microcyclic species, the number of spore types is reduced and often these rusts are restricted to a single group of host plants. Two significant rust diseases, white pine blister rust and willow rust, are described in detail below but many others attack woody plants.

CRONARTIUM RIBICOLA (WHITE PINE BLISTER RUST)

C. ribicola is native to southeastern Europe, where it co-evolved with *Pinus cembra* and related five-needle pines, and the alternate host of *Ribes*. In the early 19th century, there were several introductions of North American five-needle pines into Europe, for example *P. strobus* (eastern white pine) and *P. monticola* (western white pine). These species showed good growth and high economic potential and were planted in many countries, but it was soon realized that they were highly susceptible to *C. ribicola*.

At the end of the 19th century, *C. ribicola* was inadvertently introduced into eastern North America on *Pinus strobus* stock imported from Europe. The disease was first recorded in Massachusetts in 1898 and spread to epidemic proportions in North American forests, particularly where the climate favoured basidiospore dispersal. The large natural forests of very susceptible American five-needle pine species, and the presence of many native *Ribes* species in the forest understorey, contributed to the spread of the disease.

Symptoms

The life cycle of this rust is illustrated in **417**. Young trees and seedlings are very susceptible to infection, although obvious symptoms may not arise for three years. Needle infection occurs in spring, with red-brown flecks sometimes appearing on them, and quickly spreading into the shoots and branches. Perennial cankers form on branches and the main stem, with pycnidia developing on canker margins. A nectar containing pycnidiospores forms in the pycnidia. The nectar attracts insects which also visit other pycnidia, so exchanging pycnidiospores between opposite mating strains of the pathogen.

Orange-coloured fruiting structures (aecidia, **414**) arise annually in spring on the periphery of the canker and produce large numbers of aecidiospores. The consequent dieback of branches and the tree is associated with girdling lesions.

The aecidiospores from infected pines infect *Ribes* (**415**), where small yellow-orange uredosori (**416**) are produced on the undersides of the leaves. Uredospores (summer spores) can re-infect *Ribes* over long distances. Later in the season, rust-coloured columnar teleutosori produce teleutospores which overwinter on fallen leaves, then germinate in early spring and release basidiospores to infect needles of susceptible pines.

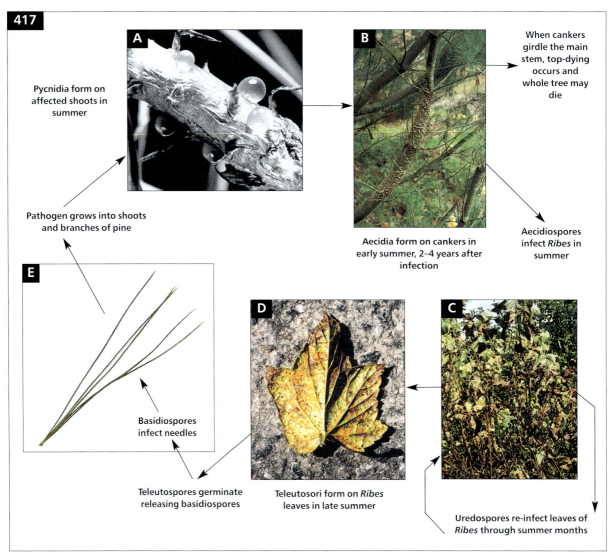

417 Life cycle of *Cronartium ribicola* on five-needle *Pinus* and *Ribes*.

418–421 Symptoms of rust diseases on angiosperm trees: **418**, yellow uredosori and orange-red teleutosori of *Melampsora epitea* on leaves of *Salix daphnoides*; **419**, dieback of this biomass willow following an attack by *M. epitea*; **420**, rust spotting on underside of a leaf of *Populus* hybrid; **421**, symptoms of *Melampsoridium betulinum* on leaves of *Betula pubescens*.

Management

During the 1930s attempts were made to eradicate *Ribes* in North American forests. However, such control is very difficult to achieve because of the persistence of *Ribes* seed in soil. Fungicide spraying programmes in the 1960s were also ineffective and unacceptable in the forest environment.

A resistance breeding programme against *C. ribicola* was instigated in the USA in 1950. The small numbers of white pine trees showing resistance to the disease were selected for breeding. Good gains have been made, despite the development of races of the pathogen able to overcome major gene resistance in some host lines. Planting resistant *Pinus strobus* lines is now recommended in integrated management schemes in the USA.

OTHER SIGNIFICANT RUST DISEASES OF GYMNOSPERMS

There are many other rust diseases of significance on gymnosperms. In North America, one of the most damaging is fusiforme rust of pines, caused by *Cronartium quercuum* forma specialis *fusiforme*. *Pinus taeda* and *P. elliottii* are particularly badly affected, although many other pines are attacked. The alternate hosts are *Quercus* spp. (oaks); certain other members of the Fagaceae are susceptible in artificial inoculations. Symptoms on pine include canker-like galls on the branches, where the aecidiospore stage is produced, ultimately resulting in girdling and dieback. Galls may expand to reach the main stem and cause death of the tree. The disease became a major problem in plantations of *P. taeda* and *P. elliottii* growing in areas with suitable conditions for infection and pathogen development, namely the presence of suitable alternate hosts, frequent periods of high relative humidity (>97%), and temperatures of 15–27°C.

Other forms of *C. quercuum* cause gall rusts on various pine species in North America.

168

WILLOW AND POPLAR RUSTS

Salix spp. (willows) and *Populus* spp. (poplars) planted for biomass production are often badly affected by rust diseases. As many as 20 species of *Melampsora*, with alternate hosts including *Larix*, *Allium*, and *Euonymus*, attack willows worldwide. *M. epitea*, the most damaging species in European biomass plantations, is divided into numerous subspecies. Poplars are also attacked by *Melampsora* species, although it is difficult to separate many of these on the basis of morphology alone.

Symptoms and life cycle

Yellow uredosori develop on leaves and soft shoots of affected plants in late spring, producing uredospores that re-infect other susceptible hosts (**418**). Infection causes distortions in leaf and shoot growth, resulting in dieback (**419**). Later in the year, rust-coloured teleutosori form on the leaves. In the following spring, the teleutospores germinate, releasing basidiospores to infect the alternate hosts on which the pycnidial and aecidial stages of the life cycle occur. Susceptible willows and poplars are infected by aecidiospores in spring.

Losses

Severe outbreaks of rust can rapidly defoliate susceptible willows and poplars, causing reduced growth and sometimes death, with the particularly high risk of severe disease occurring in monoclonal biomass plantings. Plants which are weakened by rust may become more susceptible to attack by other diseases.

Table 10 Examples of wilt disease-causing pathogens of trees

Species	Locality	Principal hosts	Figures	Comments
Ceratocystis platani	North east USA; Mediterranean countries	*Platanus* spp., esp. *P. orientalis*, *P. x hispanica*	**422, 423**	Canker stain of plane trees
Ceratocystis fagacearum	North America	*Quercus* spp., esp. red oaks	**424, 425**	Nitidulid beetles act as vectors; spread also facilitated by suckering in red oaks
Ophiostoma ulmi	Europe	*Ulmus* spp.	–	Less serious than *O. novo-ulmi* infections; 30% of infected die
Ophiostoma novo-ulmi	North America; Europe	*Ulmus* spp.	**427**	Arguably the most significant disease of plants ever recorded
Verticillium albo-atrum *Verticillium dahliae*	Worldwide	Very wide host range, especially *Acer*, *Fraxinus*, *Tilia*, *Catalpa*, *Rhus*	–	Soil-borne pathogens. Cause problems in ornamental plantings but also present in forests
Erwinia salicis	Northern Europe	*Salix alba*, *S. caprea*, *S. cinerea*, *S. fragilis*, *S. triandra*	**426**	Bacterial disease, watermark of cricket bat willow. Affected trees may die back. Reduces wood strength

Management

The pathogens adapt rapidly to the presence of resistant host cultivars, therefore effective control over long periods is difficult to achieve. Fungicides are neither effective nor economic methods for control. The selection of more-resistant host genotypes and the planting of clonal mixtures are the most promising management methods now under evaluation.

WILT DISEASES

Wilt diseases caused by various pathogens occur throughout the world (*Table 10*, **422–427**). The most important wilt of trees is Dutch elm disease; indeed, this disease is arguably the most significant plant disease ever recorded.

Symptoms

The young shoots and foliage of infected trees collapse rapidly, due to irreversible loss of turgor (**425, 427**), and the leaves later become yellow (chlorotic). These symptoms, however, may also be caused by severe root disease, so that additional symptoms are important in diagnosing vascular wilt diseases. The infection may be remote from the site of symptoms and marked disturbances occur in the water relations of the plant. The wilt is not entirely due to physical blockage of the xylem by the pathogen, since toxins are produced which are transported in the sap. Characteristic brown or green stain develops in the walls of the infected xylem (**423, 427**).

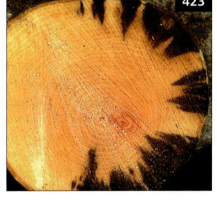

422–426 Wilt diseases of trees: **422**, mature *Platanus orientalis* dying as a result of infection with *Ceratocystis platani* (photo copyright of P. Tsopelas, NAGREF, Greece); **423**, typical staining in trunk of *P. orientalis*, caused by infection with *C. platani* (photo copyright of P. Tsopelas, NAGREF, Greece); **424**, oak wilt infection centre on *Quercus ellipsoidalis* – note the dead trees to the right, live trees to the left, and dying tree in the centre (photo copyright of J. Juzwik, USDA Forest Service); **425**, close-up of attached leaves showing symptoms of oak wilt on *Quercus ellipsoidalis* (photo copyright of J. Juzwik, USDA Forest Service); **426**, *Salix alba* var. *coerulea* showing dieback symptoms associated with *Erwinia salicis* (watermark disease).

OPHIOSTOMA NOVO-ULMI (DUTCH ELM DISEASE)

Dutch elm disease (DED) decimated *Ulmus* spp. (elm) populations in Europe, North America and western Asia in the 20th century. Outbreaks of a less-aggressive species, *Ophiostoma ulmi*, occurred in Europe from 1918 to the mid-1960s, leading to the deaths of approximately 30% of infected trees. In the UK, the incidence of the disease had fallen by 1960, apparently due to reduced virulence of the pathogen. From the early 1930s in North America, however, a particularly severe outbreak of DED was in progress, with the susceptible *Ulmus americana* suffering badly.

In the mid-1960s, a severe outbreak began in southern England, and the strain of the pathogen responsible, now known as *O. novo-ulmi*, was shown to be identical to that found in North America. The disease was imported on elm from this source and has since spread throughout almost all of the UK. It is also present from southern Scandinavia to Mediterranean regions.

The situation is further complicated by the presence of two highly aggressive strains of *O. novo-ulmi* in Europe. The North American strain, imported into western seaboard countries, is spreading eastwards, while a second strain is spreading to the west from the Caucasus. In some areas, the two co-occur and mate, giving the possibility of further highly virulent strains emerging.

Infection with *O. novo-ulmi* kills the aerial parts of elms. However, in some trees their roots may sprout, hence many hedgerow elms still exist as large shrubs or small trees. Their tops die back, however, after several years of growth, once they reach a suitable size for the beetle vector to feed on them. Elms were large, significant components of the landscape, therefore marked changes in the amenity value of affected areas has occurred.

Life cycle

In Europe the disease is transmitted from tree to tree (427) by several species of elm bark beetles (*Scolytus scolytus*, *S. multistriatus*, and *S. laevis*). The major factor in long-distance spread is the transport of infected logs within and between countries.

The beetles overwinter in breeding galleries in the bark of dying or dead trees and the fungus develops saprotrophically in this bark. When adult beetles emerge in spring, the sticky spores from coremia, formed by the fungus in the galleries, become attached to the insects. Immediately after emergence, the beetles begin maturation feeding on bark in the upper crowns of nearby elms. Spores from the beetles' bodies infect the feeding wounds in the bark, germinate, and intrude into xylem vessels of the elm. The pathogen rapidly spreads in the xylem sap and typical wilt symptoms arise in the elm. Tyloses form in the affected xylem vessels when the plasmalemma of adjacent parenchyma cells expands into the vessel lumens. The bark of killed trees provides the breeding substrate for the next generation of beetles.

Management

Control of the beetle vectors has been attempted (by destroying infected wood, application of contact or systemic insecticides, and the use of wasps parasitic on *Scolytus*) but these measures have been ineffective. Injection of systemic fungicides, such as thiabendazole or the triazole fungicides propiconazole and tebuconizole, into elms can provide 30–40% protection, depending on the size of the tree. Such injections are widely used for Dutch elm disease control (and control of other wilt diseases) in Europe and North America. Treatment is very expensive, however, and, as it is required every year, is worthwhile only on trees with high amenity value.

Elm resistance breeding programmes in the USA and The Netherlands attempt to exploit the natural resistance found in certain Asian species of *Ulmus*. The most notable cultivar released to date is Sapporo Autumn Gold, an F_1 hybrid of *U. pumila* x *U. japonica*. The apparent greater resistance of some elms, for example *U. glabra* and *U. laevis*, to the disease reflects the lower desirability of their bark to feeding beetles.

427 Life cycle of Dutch elm disease.

VERTICILLIUM WILT

Many species of woody flowering plants, including forest trees, tree crops, and shrubs, are susceptible to wilt caused by species of *Verticillium*. In most cases, the wilt is likely to be caused by *V. dahliae*, although in some instances, other species of the pathogen may be the causal agent. The disease is a particular problem on *Olea* (olives), *Acer* spp. (maples), certain stone fruits (*Prunus* spp.), and *Pistachia vera* (pistachio). Maples in amenity plantings can be very badly affected under conditions conducive to disease development.

Infection leads to the development of typical wilt symptoms, as described above.

Life cycle

V. dahliae may survive for at least 10 years in soil as microsclerotia in dead and dying plant tissues; weed species also provide reservoirs of infection that can maintain viability of the pathogen over many years. When roots grow, substances are exuded into the soil; these chemicals can stimulate germination of *V. dahliae* microsclerotia, and the germinating hyphae are attracted towards the source of the exudates. On contact with the surface of a fine root, the hyphae penetrate the host, growing into the vascular cylinder and entering the xylem. Production of specialized conidiospores (bud cells) in the xylem vessels enables the pathogen to spread systemically very rapidly in the host. Symptoms are generated as described above.

Oak wilt

Death of *Quercus* spp. (oaks) through wilting was noted in the North Central States of the USA in the mid-19th century, although the causal agent was not discovered until the mid-1940s. Oak wilt is caused by the fungus *Ceratocystis fagacearum*, and is known to kill about 20 species of oak. Since the first description of the disease, it has spread into Texas; it is, however, confined to the eastern side of the Rocky Mountains. Red oaks (sub-genus *Erythrobalanus*) are particularly susceptible and die within one to two months of the first symptoms becoming apparent (**424, 425**). In contrast, white oaks (sub-genus *Leucobalanus*) may live for many years with the infection, taking up to seven years to die, although cases are known where the trees have remained alive for at least 20 years.

Symptoms vary between red and white oaks, and with geographical region. A common symptom is the formation of grey to dark brown streaks in the xylem vessels. Susceptible red oaks include *Q. rubra* (red oak), *Q. palustris* (pin oak), *Q. tinctoria* (black oak), *Q. coccinea* (scarlet oak), *Q. imbricata* (shingle oak), and *Q. nigra* (blackjack oak). In the *Leucobalanus* sub-genus, *Q. alba* (white oak), *Q. bicolor* (swamp white oak), *Q. macrocarpa* (bur oak), *Q. pinoides*, and *Q. obtusifolia* (post oak) are damaged to varying degrees.

Red oaks

In the North Central States, the first symptoms are the development of necrotic patches on the leaves in summer, particularly at the leaf tip or on the edges. Leaves discolour, becoming a dull olive-green, through red-brown to bronze (**425**). Premature abscission results; leaves on the ground can show all gradations of colour listed above, and are often curled up at the edges. Dieback is rapid, from the top downwards, and the whole tree can die quickly (**424**). Close examination of the main trunk may show vertical cracks, covering the developing spore mats, which may be 5–18 cm or more in length. Spore mats are grey–black felt-like structures that produce a fruity odour which attracts insects, birds, and squirrels; sometimes squirrels will gnaw through the bark over spore mats to gain access to the sugary exudate. As many red oaks commonly form clonal clumps through suckering, large patches of trees may appear to die within woodlands.

White oaks

When white oaks become infected with wilt, one or a few branching systems may develop symptoms, most obviously in mid- to late summer. Dieback varies greatly between trees, and leaf disease symptoms are rather similar to those found in infected red oaks.

Life cycle

Human activity plays a major role in disease spread. Beetles that transmit the pathogen are attracted to damaged trees; any pruning activity or damage through building works can lead to infection. Transmission of oak wilt by oak bark beetles – *Pseudopityophthorus minutissimus* and *P. pruinosus* – is similar to that of Dutch elm disease by *Scolytus* species. Beetles are attracted to recently dead oaks, and lay eggs. The larvae feed in the phloem tissues and pupate; the newly emerging adults may pick up spores of *C. fagacearum*, if the disease killed the trees. Maturation feeding of the newly emerged adults in the crown of healthy trees then transmits the infection to new hosts. In the North Central States of the USA, the pathogen is also vectored by a range of oak sap beetle species, including *Carpophilus*, *Colopterus*, *Cryptarcha*, *Epurea*, and *Glischrochilus*. These beetles are attracted to active spore mats, as described above, and pick up spores during feeding on the sugary exudate. When visiting wounds, including pruning wounds on healthy trees, the infection is transmitted by contact.

On infection, the spores enter the vascular system of the tree and spread systemically. Wilting is probably due both to vessel occlusion through tylosis formation, and toxin production by *C. fagacearum*. Within eight weeks of defoliation, spore mats are formed under the bark on the main branches and trunk of the infected tree. As the fungal material builds up, pressure is exerted on the dead bark, which cracks, allowing entry by sap beetles.

Management

Several management protocols have been instigated in the central states of the USA, where oak wilt is most serious. As the disease attacks host trees in forests, woodlands, parks, and gardens, there is an education policy that aims to inform the public of the symptoms and infection biology of the disease, plus likely methods of management. Methods adopted include surveying for the disease, management to reduce wounding or to treat wounds with paint within minutes of damage occurring, root graft cutting to prevent physical transmission, and injection of the fungicide propiconazole (see management of Dutch elm disease).

VIRUS DISEASES

Virus diseases of plants are extremely important worldwide, causing huge economic losses. Most plants become infected by viruses, although some may have little apparent effect on plant growth. Some examples in trees (**428–431**) include poplar mosaic carlavirus, apple mosaic ilaravirus, and cherry leaf roll nepovirus, while many other virus-like infections have been noted.

Importance

Poplar mosaic carlavirus can cause 30% loss in increment in certain *Populus* (poplar) cultivars. Very little information is available in relation to losses to virus infections in other trees, with most coming from work on fruit trees.

Viruses are generally not host specific, infecting a wide range of unrelated hosts. Forest trees, therefore, may become reservoirs of infection for other crops. For example, tobacco ringspot virus was found in *Cupressus arizonica* (Arizona cypress), tobacco necrosis virus has been isolated from *Larix decidua* (European larch), and strawberry latent ringspot was found in species of *Aesculus*.

Virus diseases may appear unimportant in forestry, but this assumption probably reflects a lack of knowledge about tree viruses and their impact on forest ecosystems.

428–431 Symptoms of virus infections on trees (photos copyright of I.E. Cooper, CEH, Oxford): **428**, leaf of *Populus* showing symptoms of chlorosis caused by poplar mosaic carlavirus infection; **429**, *Betula* leaves with ring spot symptoms resulting from apple mosaic ilaravirus infection; **430**, walnut blackline, a graft incompatibility problem associated with cherry leaf roll nepovirus; **431**, dwarfing of *Betula* sapling (left) inoculated with cherry leaf roll nepovirus. Control sapling is on the right.

Symptoms

Different symptoms may occur with the same type of virus on different tree hosts. Generalized symptoms are similar to many of those described earlier in this chapter, and include chlorosis (sometimes known as 'mosaics', where mottling occurs over the leaf lamina), yellowing of the veins, yellow circles or necroses on the leaves, and patches of killed bark. Increased (hyperplasia) or decreased (hypoplasia) cell division may also occur, or tree growth may cease (atrophy). Viral infections can cause distortions of leaves or shoots, and damage to underlying woody tissues.

EXAMPLES OF VIRUSES IN FOREST TREES

Virus-like particles in gymnosperms were not recognized until the 1960s, and many of these agents have not been extensively characterized. In angiosperm trees, however, many virus diseases have been reported (Cooper, 1993). Poplar mosaic carlavirus affects poplars (apparently the only hosts) of the *Populus deltoides* group in Europe, North America, and Japan. Foliar symptoms include spotting, mottling, and mosaic (**428**); leaves may curl and develop small growths on petioles.

Cherry leaf roll nepovirus occurs on several hosts, including cherry (*Prunus avium* and *P. padus*), and species of *Betula* (birch), *Fagus* (beech), *Fraxinus* (ash), *Juglans* (walnut), and *Ulmus* (elm). The symptoms are leaf rolling, diffuse mottling of leaves, chlorotic spots or lines, rings, and patches. Elm mosaic virus is common in shade elms in North America, causing dieback of upper branches, sparse foliage, small leaves with chlorotic ring patterns, and small overgrowths under the leaves in the interveinal regions.

Life cycle

Mechanical transmission is probably very rare in nature, but occurs in normal horticultural practices via contaminated tools or during grafting. Otherwise, to gain entry to host plants, viruses rely on vectors. The most important vectors are insects (particularly aphids), leaf hoppers, whiteflies, and mealy bugs. Other arthropod vectors include mites.

Virus particles comprise an infectious nucleic acid encapsulated within a protective protein coat. In most plant viruses the genome is RNA, although a few groups are DNA-based. The RNA or DNA may be single- or double-stranded. Some plant viruses contain additional satellite strands of genomic material, which may modify the infection process. Once inside the host plant, viruses replicate within infected cells, producing large numbers of new virus particles, which may then be transmitted to uninfected hosts by vectors.

Management

Chemotherapy and/or thermotherapy, or meristem culture, are used to eliminate viruses from horticultural crops. Control in trees, other than fruit trees, is generally not practical. However, genetic manipulation by inserting DNA encoding viral coat proteins or satellite RNA sequences into the plant genome has been tested against poplar mosaic carlavirus in *Populus* spp. grown for biomass production. In any case, however, treated trees would be exposed to re-infection once they are planted out. These various control methods are therefore not suitable for forest trees, which have very long rotation times.

PARASITIC PLANTS

Parasitism of plants by other plants is found in all forested parts of the world, from the boreal regions to the tropics. The term 'mistletoe' encompasses over 1,300 species of hemiparasitic flowering plants which abstract water and minerals from a host plant via specialized haustoria in the host's vascular system, yet are mostly photosynthetic and therefore able to synthesize sugars. The most important mistletoes attacking forest trees are *Arceuthobium* spp. (dwarf mistletoes, **432**) and various species of *Viscum* (**433–435**), *Amyema* (**436**), *Loranthus*, *Phoradendron*, and *Psittacanthus* (leafy mistletoes). Mistletoes, however, have important ecological roles in forest ecosystems, relying on animals for pollination and dispersal, and forming an important habitat and food plant for many forest organisms. In addition to those species with an obvious aerial parasitic habit, several species of tree are hemiparasitic, relying on the roots of adjacent trees for sustenance. Species of *Santalum* (sandalwood) are important examples of this habit (**437**); *S. album* wood has been used for centuries to produce fragrances.

432–437 Mistletoe infections of trees: **432**, spring shoots of *Arceuthobium oxycedri* emerging from branch of *Juniperus oxycedrus*; **433**, swelling of *Populus* host branch at point of infection by *Viscum album*; **434**, multiple *V. album* infections in the crown of *Salix alba*; **435**, multiple *V. abietis* infections in crown of *Abies cephalonica*; **436**, *Amyema* (probably *A. pendulum*) in a roadside *Eucalyptus* sp., South Australia; **437**, parasitic sandalwood (quandong; *Santalum acuminatum*), probably attached to adjacent *Eucalyptus* sp., South Australia (photo copyright of Mark Nurmela, Mount Barker, South Australia).

DWARF MISTLETOES

Forty-four species of *Arceuthobium* are found in the northern hemisphere, mostly in North America, where they cause major losses on conifers. Identification is based on their morphology, cytology, chemistry, and host specificity (Hawksworth and Wiens, 1996). Some of the most significant species are: *A. abietinum*, which attacks species of *Abies* (fir) in western North America; *A. californicum*, parasitic on species of *Pinus* (pine) in California and Oregon; *A. douglasii*, which attacks *Pseudotsuga menziesii* (Douglas fir) in western North America; and *A. laricis*, occurring on *Larix* spp. (larch) in Oregon, Washington, and British Columbia. There is a single species recorded in the Old World, *A. oxycedri*, which is common on *Juniperus oxycedrus* in Mediterranean regions and in northern India (**432**).

Life cycle

Seed is discharged explosively up to 10 m from the mistletoe plant and sticks to needles on the surrounding trees. On wetting by rain, the seeds germinate and penetrate the host tree. Mistletoes typically flower from one to three years after infecting a host. Male and female flowers are borne on separate dwarf mistletoe plants, which produce new shoots and flowers annually. The presence of witches' brooms on a tree is a typical symptom of infection, together with the formation of cankers, crown dieback, and reduction in tree growth rate.

Losses

In North America, damage by dwarf mistletoes is second in importance to that caused by decay. Approximately 5.5 million ha of the conifer forests of California, Oregon, and Washington State are affected by *Arceuthobium*. Losses in timber in the USA have been estimated at 11.5 million m^3 per annum, with considerable additional impacts on forest ecology and stand dynamics.

Management

Current methods aim to confine infection at an acceptable level, rather than eliminating infections altogether. The use of non-susceptible conifer species and forest clearings as buffer zones is proving reasonably successful. In areas of high amenity value, intensive control methods such as pruning, or fertilizing to increase host tree vigour, can be useful.

LEAFY MISTLETOES

Viscum album occurs in Europe, attacking a range of broadleaved trees, including *Populus* x *euamericana* (black poplars, **433**), *Malus domestica* (apple), and species of *Salix* (willow, **434**) and *Tilia* (lime). *V. abietis* occurs in Spain, France, Greece, and Romania; it primarily attacks *Abies* spp. (fir, **435**) but may also affect species of *Larix* (larch) and *Picea* (spruce). *V. laxum* occurs on *Pinus* spp. such as *P. nigra* in southern Europe.

V. album and related species are dioecious, with the female plants producing white berries containing sticky seeds. Birds eat the berries and the seeds stick to their beaks; the birds vigorously rub their beaks on tree branches to remove the seeds, which are then deposited there. At germination, the radicle of the mistletoe penetrates the bark of a host tree to reach the secondary phloem and sapwood, where it forms absorptive haustoria (sinkers). Mistletoe is perennial, and its continued uptake of nutrients and water from the host has a weakening effect on the tree. However, the loss in yield of affected trees can be offset by the sale of mistletoe for Christmas decorations, and the deliberate cultivation of mistletoe on old apple trees is common in parts of northern Europe.

Several species of *Amyema* (family Loranthaceae) are common on species of *Eucalyptus*, *Acacia*, *Leptospermum*, and *Casuarina* in Australasia, including New Guinea. These leafy mistletoes have similar life cycles to *V. album*, but have distinctly different adaptations to the varying hosts and environmental conditions found in Australasia. *A. pendulum* (**436**), for example, is common on eucalyptus trees in New South Wales and South Australia. *Amyema* and other leafy mistletoes are recognized as important nesting and feeding habitats for Australian birds.

ACKNOWLEDGEMENTS

The following colleagues kindly gave permission for the use of illustrations: Ellen Goheen and Jenny Juzwik (USDA Forest Service), Eric Boa (CABI Bioscience), C.S. Hodges (University of Gainesville), Don Barrett (formerly of the Oxford Forestry Institute), Ian Cooper (CEH Oxford), Gary Griffin (Virginia Polytechnic Institute and State University), W.J.A. Bodles (University of Aberdeen), Wiesław Mułenko (Marie-Curie-Sklodowska University, Lublin, Poland), Stan Murray (formerly of the University of Aberdeen), Stan Thompson (Forres, Morayshire), Mark Nurmela (Mount Barker, South Australia), Richard Robinson (Conservation and Land Management Department, Western Australia), and Paolo Capretti (Universita degli Studi, Firenze).

CHAPTER 9

Insect pests of some important forest trees

Claire Ozanne

INTRODUCTION

Insects feeding on trees are described as pests when they cause damage that is unacceptable to humans either in biological or economic terms. Across the world's forests insects are estimated to take between 5 and 15% of tree foliage annually (Speight and Wainhouse, 1989), although some studies have found insect herbivores to have a lower impact (0.7% in mature pine forest; Larsson and Tenow, 1980), while others record much higher losses (up to 20% in Australian rainforest; Lowman, 1984).

Most insect pests cause damage to trees when they feed, either as adults or juveniles, and may do so directly or indirectly. Direct damage results when part of the tree is removed – leaves, fruit, bark, wood, or root – and the material is usually consumed by the insect. Such feeding can reduce foliar or root area (and therefore productivity of the tree), may interrupt its nutrient and water flow, and can also cause changes to the form of the tree, such as bending of its shoots, branches, and even the trunk. Feeding on the seed and fruit reduces the viability of seed orchards and affects natural regeneration. Indirect damage results from the introduction of a disease-causing organism, such as a fungus or bacterium, into the tree during the feeding or egg laying process. Several insects have complex symbiotic relationships with various micro-organisms, which can cause disease when inoculated by the insect into a susceptible host tree (**438**).

438 Bluestain fungus staining the wood of *Picea sitchensis* (Sitka spruce). The fungus was transmitted by *Trypodendron* sp. (Coleoptera: Scolytidae) and the tree was also attacked by *Hylocoetus* sp. (Coleoptera: Lymexylonidae).

Insect pests can be divided into several categories based on the way in which they feed and therefore the damage that they cause to the tree. These categories include: suckers or sap feeders; bark and shoot borers, and chewers (bark chewers and defoliators). This chapter includes examples of these main groups and covers a range of temperate and tropical host tree species.

SAP FEEDERS
ELATOBIUM ABIETINUM WALKER (HOMOPTERA: APHIDINAE, APHIDIDAE) – GREEN SPRUCE APHID
Description
The adult and juvenile aphids are green and pear-shaped with dull red eyes (**439**) and distinct tube-like structures or cornicles (thought to secrete wax) on the abdomen. The juveniles or nymphs are wingless, and normally measure around 0.1–0.15 cm in length. The adults may be winged or wingless, and are approximately 0.2 cm in length. The eggs are oval and are yellowish-red to brown or black (Ministry of Forestry Canada, Day *et al.*, 1998).

Host trees
All species of *Picea* (spruce) that are currently in cultivation, with *P. sitchensis*, *P. pungens*, *P. engelmanni*, and *P. alba* all seriously damaged by aphid attack (Koot, 1991).

Distribution
The aphid is widespread in the UK, with a continuous distribution across northern France, Denmark, Germany, Switzerland, and Austria. The aphid also occurs in small areas of Scandinavia, Latvia, and Iceland and has a limited distribution in Tasmania, New Zealand, and the east and west coasts of North America (Day *et al.*, 1998).

Impacts/ecology
The aphids feed on the underside of needles on the mesophyll cells; mature needles are preferred and the aphids are sensitive to variations in their terpene composition. The first signs of aphid feeding are yellow patches on the needles (**440**), which then turn yellow and brown and may be lost during the summer (Koot, 1991). Outbreaks are recognized when the population reaches 0.5 aphids per needle in the summer population peak. The impact on height growth in *Picea sitchensis* (Sitka spruce) has been found to be a 10–30% reduction following severe spring/early summer defoliation, and a narrowing in stem diameter is evident both in the year following defoliation and even up to seven to eight years on. Fertilizer and SO_2 and NO_2 treatments render trees more susceptible to attack, and the response to elevated CO_2 is currently being investigated (Day *et al.*, 1998).

439 Apterous (wingless) form of *Elatobium abietinum* (green spruce aphid) on *Picea sitchensis* (Sitka spruce) needle.

440 Damage caused by *Elatobium abietinum* to the foliage of *Picea sitchensis* – note yellowing of the needles.

Control

Epidemics are often reduced by weather changes such as the onset of cool temperatures. Coccinellid beetles and hemerobiids (brown lacewings), and other such predators are most likely to have a major impact on aphid densities. However, parasitoids (parasites that develop within and kill the host – usually from the orders Hymenoptera, parasitoid wasps, or Diptera) also occur in aphid populations (Day *et al.*, 1998). Chemical control is not viable in plantations but may be used in nurseries.

CINARA CUPRESSI BUCKTON (HOMOPTERA: APHIDIDAE, LACHNINAE) – CYPRESS APHID
Description

The adults and juveniles are brownish, soft-bodied aphids, often covered in a grey waxy coating (**441**), and the adults are approximately 2.4 mm long (O'Neil, 1998). *Cinara cupressi* is a member of a species complex which includes *C. cupressivora*, *C. sabinae*, *C. cupressa*, and an as yet unnamed species (Watson *et al.*, 1999).

Host trees

The species complex feeds on trees in the Cupressaceae (cypress family) (Watson *et al.*, 1999). These include: *Cupressus macrocarpa* (Monterey cypress) and other species; *Juniperus scopulorum, J. virginiana*, and *J. sabinae* (junipers); *Thuja occidentialis* (white cedar) and *T. plicata* (western red cedar); *Chamaecyparis lawsoniana* (Lawson cypress); *Widdringtonia* spp.; and *Callitis calcarata* (*C. cupressivora*).

Distribution

Cinara cupressi is found in Europe (Germany, The Netherlands, and the UK). It also occurs in Sri Lanka, the USA (California, Pennsylvania, Arizona, Colorado, and Utah) and in Ontario, Canada. Aphids also damage trees in British Colombia, Israel, Poland, Slovakia, and Lithuania; but these attacks are likely to be caused by *C. cupressivora*. The latter species is also found in Kenya, Rwanda, South Africa, Malawi, Mauritius, Morocco, Zimbabwe, Syria, Jordan, Yemen, Burundi, Colombia, France, Greece, Italy, and Turkey.

Impacts/ecology

Aphids pierce the bark of tree twigs and feed on the phloem sap, resulting in death of the twig and leading to yellowing of the foliage, especially in tropical dry seasons (**442**; O'Neil, 1998). Ants often tend the aphids and move them around the tree, so increasing

441 *Cinara cupressi* (cypress aphid) on the young shoots of *Cupressus macrocarpa* (Monterey cypress).

442 Damage to *Cupressus* sp. (cypress) in Malawi caused by *Cinara*.

aphid spread. Sooty moulds commonly grow on the honeydew exuded by the aphids, covering the leaves and reducing the photosynthetic capacity of the tree. If infestations are severe, feeding may cause tree mortality. Loss of productivity from aphid infestations in East and Central Africa is estimated at US$13.5 million per annum (Wingfield and Day, 2002).

Control
The parasitoid wasp *Pauesia juniperorum* could be used as a control agent in the tropics as it will only parasitize aphids in the sub-family *Lachninae*. In France and South Africa, *Pauesia* species introduced to control similar aphids began to control populations within two years, and studies by IIBC (International Institute of Biological Control, unpublished data) have shown that in the laboratory, *Pauesia juniperorum* can cause substantial mortality of *Cinara cupressi* (O'Neil, 1998). The aphid may also be controlled by early applications of chemical agents containing Pirimicarb.

HETEROPSYLLA CUBANA CRAWFORD (HEMIPTERA: PSYLLIDAE) – LEUCAENA PSYLLID
Description
The adult psyllid (443) is yellow, winged, and approximately 2 mm long (Hertel, 1998). Nymphs

443 Winged (alate) adults of *Heteropsylla cubana* (leucaena psyllid) on young shoots of *Leucaena*.

are similar to the adults, although smaller in size. Eggs are laid between the new leaves on young host shoot tips and the insect can complete its life cycle in 10–15 days when conditions are optimum.

Host trees
Found only on *Leucaena* spp. (leucaenas).

Distribution
The psyllid is native to Central America but the pest began to spread towards the west in the early 1980s and is now pantropical in distribution.

Impacts/ecology
Both adults and nymphs feed on sap in young buds, shoots, leaves, and flowers of *Leucaena* trees. After an attack, newly emerged leaves will not fully unfurl, and shoots are subject to wilting; susceptible varieties may lose their leaves and, in severe cases, die (Geiger and Gutierrez, 2000). A sooty mould often grows on the honeydew produced by the psyllids, covering the leaves and reducing photosynthesis. In the Asia Pacific region the psyllid caused great damage within the first two years of arrival (Heydon and Affonso, 1991), with losses in potential productivity of up to 33% (Oka, 1989).

Control
The incidence of the psyllid may be reduced by wide spacing of newly planted trees to reduce humidity in the stand. Chemical control has proven ineffective against the psyllid, although the use of dimethoate has been recommended in some areas. Biological control agents such as the parasitoids *Psyllaephagus yaseeni* Noyes (Encyrtidae) and *Tamarixia leucaenae* Boucek (Eulophidae), and the coccinellid beetle predator *Curinus coeruleus*, have had some efficacy in controlling populations in Hawaii and Indonesia (Nakahara *et al.*, 1987; Mangoendihardjo and Wagiman, 1989; Wagiman *et al.*, 1989). However, no highly effective agent has yet been found.

ADELGES TSUGAE ANNAND (HEMIPTERA: ADELGIDAE) – HEMLOCK WOOLLY ADELGID

Description

The adults are small reddish-purple insects, approximately 1 mm in length, and are covered in a white waxy secretion (**444**). There are winged (alate) and wingless (apterous) forms of adult; the latter are slightly smaller in size. The nymphs are similar to small wingless adults. The adults lay around 50–100 orange-brown to black eggs in a batch, and up to 300 in a lifetime. The eggs are covered in white wax.

Host trees

The primary hosts for the sexual phase of the insect's life cycle are *Picea* trees (spruces) – this part of the life cycle has not yet been recorded to have been completed successfully in northern America, where the insect has recently invaded. The secondary hosts in northern America are *Tsuga canadensis* (eastern hemlock) and *Tsuga caroliniana* (ornamental Carolina hemlock).

Distribution

The adelgid is native to Japan and China but ranges in North America from Nova Scotia to Minnesota, and southwards to northern Alabama (Baumgras *et al.*, 1999). It is currently found in approximately 25% of *Tusga canadensis* forests in North America (Zilahi-Balogh *et al.*, 2002).

Impacts/ecology

The adelgid feeds on the tissue at the base of the needles, causing nutrient loss and dehydration (McClure, 1996). This feeding, together with the injection of the sheath of saliva from around the thread-like mouthparts of the adelgid, results in needle loss within a few months of infestation. Subsequently, the buds, twigs, and branches die and eventually the trees may be killed within three to four years if the attack is left uncontrolled (**445**; Young *et al.*, 1995; McClure, 2005). The insects are dispersed by wind, birds, and mammals (McClure, 1990), and populations can spread up to 25 km per year (Yorks *et al.*, 1999).

Control

The current control strategies include early detection and the removal and destruction of infected trees. The promotion of good silvicultural practices designed to prevent tree stress may reduce the rate of attack. Trees can be sprayed with horticultural oils and chemicals such as Imidacloprid; this treatment is most effective in urban situations (McClure, 2005). Biological control agents such as the predatory coccinellid beetle *Sasajiscymnus tsugae* (Japanese lady beetle; Cheah *et al.*, 2004) and the oribatid mite *Diapterobates humeralis* have provided some control in plantations. The efficacy of a predatory derodontid beetle, *Laricobius nigrinus*, is currently being investigated (Lamb *et al.*, 2005).

444 White waxy deposits on the needles of *Tsuga* sp. (hemlock) caused by secretions from the adults of *Adelges tsugae* (hemlock woolly adelgid). (Photo copyright of Christopher R Webster, Michigan Technological University, USA.)

445 Stand of *Tsuga* sp. (hemlock) heavily defoliated by *Adelges tsugae* (hemlock woolly adelgid). (Photo copyright of Christopher R Webster, Michigan Technological University, USA.)

BARK AND STEM BORERS
DENDROCTONUS MICANS KUGELANN
(COLEOPTERA: SCOLYTIDAE) – GREAT
EUROPEAN SPRUCE BARK BEETLE

Description

This is a large black cylindrical beetle 6–9 mm in length, with a covering of orange-brown hairs (446). The larvae, which feed in aggregations, are typical in form for this family, and legless with pale cream bodies and brown head capsules (447). The life cycle of *D. micans* is not as predictable as other related pest species of bark beetle, and adults exhibit pre-emergence mating (King and Fielding, 1989).

Host trees

The beetles will feed on a range of conifers, including *Abies* (firs), *Larix* (larches), *Picea* (spruces), *Pinus* (pines), and *Pseudotsuga* (Douglas fir and other species). *Picea abies* (Norway spruce) is the main host in Europe, but infestations of *P. orientalis* (oriental spruce) and *P. sitchensis* (Sitka spruce) also occur. In eastern Eurasia the main host tree is *P. orientalis*.

Distribution

D. micans is the principle Eurasian species in this genus. Its range stretches from northern Japan and Sakhalin Island across northern and western Europe, including Austria, Belgium, the Czech and Slovak republics, Denmark, Finland, France, Germany, The Netherlands, Sweden, Romania, Turkey, the UK, the former USSR, and former Yugoslavia.

Impacts/ecology

D. micans breeds in living and healthy trees, and the larvae feed gregariously in order to overcome tree defences (King and Fielding, 1989). An attack is, however, more common when trees have been damaged or are weakened in any way, and the degree of bark lignosuberization (impregnation with a complex of carbohydrates and fatty acids) is known to affect susceptibility. The larvae excavate galleries under the bark, destroying the adjacent cambium; several attacks, perhaps over subsequent years, may result in ring barking the tree and causing death (Forestry Commission, 2002). Resin tubes and resin bleeding from tree stems characterize a larval attack (448). The fertilized females disperse by

446 Adult of *Dendroctonus micans* (great European spruce bark beetle) – note its covering of orange-brown hairs. (Photo copyright of Forestry Commission, UK.)

447 *Dendroctonus micans* legless larvae feeding in aggregation – note the contrast between their creamy-white bodies and brown heads. (Photo copyright of Forestry Commission, UK.)

448 Damage to *Larix* sp. (larch) caused by *D. micans* – note the resin bleed from its trunk.

flight (21–23°C threshold) or by walking to neighbouring trees, and epidemics are most frequently associated with the leading edge of an expanding population.

Control

A survey of beetle distribution, sanitation felling of infected trees, and the creation of quarantine or protected zones from which timber cannot be moved help to limit the spread of an infestation. Biological control with the predatory beetle *Rhizophagus grandis* (Gyllenhal Rhizophagidae) is also highly effective. In the UK, plantation control by *R. grandis* is predicted to reduce losses caused by this bark beetle from 10% to 1% (Forestry Commission, 2003).

IPS TYPOGRAPHUS LINNAEUS (COLEOPTERA: SCOLYTIDAE) – EIGHT-SPINED SPRUCE BARK BEETLE (LARGER EUROPEAN SPRUCE BARK BEETLE OR SPRUCE ENGRAVER)

Description

This is a moderate- to large-sized and cylindrical brown beetle (4.2–5.5 mm), with four spines on the lateral edge of each wing case (elytron) in males (eight in total). The anterior part of the pronotum (plate covering the upper surface of the first segment on the thorax) is rough and the body is fringed with yellow hairs (Whittle and Anderson, 1985). The larvae are white and legless, with a light-brown head capsule. They grow to about 5 mm in length.

Host trees

The main and preferred host is *Picea abies* (Norway spruce), but the beetle will also attack *Larix* (larches), *Abies* (firs), and *Pinus* (pines).

Distribution

It occurs in most of Continental Europe and northern Asia, including China, Japan, Korea, and Russia's far east.

Impacts/ecology

This beetle is an aggressive pest (Byers, 1989). Many beetles cooperate to attack single trees within a stand, and these trees are frequently killed. Such cooperation is necessary to overcome the resin defences of the tree, and is achieved using an aggregation chemical or pheromone (methyl butenol and *cis*-verbenol), which is produced by feeding males (Bakke and Strand, 1981; Byers, 1989). Galleries are excavated in the thin phloem layer situated 2–4 mm under the host bark (**449**), and the sapstain fungus *Ceratocystis polonica* is introduced (**450**; Webber and Eyre, 2003). The host tree is killed (**451**) by a combination of tissue damage by the beetles and blocking of the phloem by the fungus.

449 De-barked trunk showing the galleries created by *Ips* sp. (spruce bark beetle) larvae.

450 Cross-cut log showing evidence of sapstain fungus transmitted to the host tree by *Ips* sp.

Control

Sanitation felling may be used to limit the spread of this beetle, but chemical control is not considered to be economically viable. Pheromone lures containing the aggregation pheromones are used to monitor populations and to trap beetles. A wide range of natural enemies is currently being investigated for their effectiveness in control, such as the predatory clerid beetle *Thanasimus formicarius*, the hymenopteran parasitoids *Dendrosoter middendorffi*, *Coeloides bostrichorum*, *Tomicobia seitneri*, and *Roptrocerus xylophagorum*, and the parasitoid wasp *Rhopalicus tutela* (Morrisey, 1996; Gregoire, 2003).

AGRILUS PLANIPENNIS FAIRMAIRE (COLEOPTERA: BUPRESTIDAE) – EMERALD ASH BORER

Description

The adult beetles are metallic green in colour with an elongate, wedge-shaped body ranging in size from 7.5 to 14 mm long and 3.0 to 3.4 mm wide (**452**). The beetles have a ridge along the front margin of the wing cases and normally have black, indented or kidney-shaped eyes. The mature larvae are long and thin (26–32 mm long), cream in colour, with partially hidden brown head capsules and brown horns on the last abdominal segment (McCullough and Katovich, 2004; Global Invasive Species Database, 2005).

Host trees

In its native range the beetle can be found in *Fraxinus chinensis*, *F. rhynchophylla*, *F. mandshurica* var. *japonica* (ashes), *Ulmus davidiana* var. *japonica* (an elm), *Juglans mandshurica* var. *sieboldiana* (a walnut), and *Pterocarya rhoifolia* (a wingnut) (Haack *et al.*, 2002; Nomura, 2002). In North America it has been found attacking *Fraxinus pennsylvanica* (red-green ash), *F. americana* (white ash), and *F. nigra* (black ash), as well as several other ash species and cultivars (MacFarlane and Meyer, 2005). It is predicted that most ash varieties will be susceptible.

Distribution

A. planipennis has a native range stretching from the far east of Russia across China, Korea, Taiwan, and Japan. In 2002 it was discovered in North America and has now spread to Michigan, Ohio, Indiana, Maryland, and Virginia, and to Ontario, Canada (MacFarlane and Meyer, 2005).

451 *Pinus* sp. (pine) trees attacked and killed by *Ips* sp.

452 Adult beetle of *Agrilus planipennis* (emerald ash borer). (Photo copyright of Andrew J Storer, Michigan Technological University, USA.)

Impacts/ecology

The adults feed on tree leaves but rarely cause significant damage. The larvae burrow through the bark and feed on the cambium layer and newly forming phloem and xylem. This activity produces snaking tunnels, packed with frass (faecal pellets, **453**). These galleries can extend for up to 50 cm and interfere with the transport of water and nutrients; the bark splits, affected portions of the tree die and, after three to four years of infestation, a tree may be killed. Since ash species are widely distributed in North American forests, emerald ash borer is likely to substantially change forest diversity and dynamics. Significant economic effects are also predicted, as ash trees provide a range of important wood products and are often grown in urban areas where infestations will need intensive management (MacFarlane and Meyer, 2005).

Control

Limiting spread of the beetle over long distances by the implementation of quarantine zones and restricting the movement of trees and wood products can be effective. Chemical control may be carried out by topical application of trunk and foliage sprays such as Cyfluthrin, but the timing is crucial, as it must coincide with the emergence of the adults. Systemic chemical application using soil drenching (e.g. Imidacloprid) or injection (e.g. Bidrin) may be used to kill larvae already present in the tree (Smitley and McCullough, 2004). Biological control is currently limited to trunk spraying with fungal spores of *Beauvaria bassiana*, which are reported to cause approximately 50% mortality in emerging adults (Smitley and McCullough, 2004).

PHORACANTHA SEMIPUNCTATA FABRICIUS (COLEOPTERA: CERAMBYCIDAE) – EUCALYPTUS LONGHORN BORER

Description

The adult beetle is 2.5–3 cm in length, and a shiny dark brown to black in colour. A buff-coloured, zigzag patch crosses the elytra (wing cases), the anterior portions are covered by small depressions (punctate), and the elytral tips have a buff-coloured spot (**454**). As is characteristic of this beetle family, the antennae are as long as or longer than the body (longer in the males). The larvae are cream-coloured with a retracted head capsule and effectively legless. They can grow to 3 cm in length (Bain, 1976).

Host trees

Many species of *Eucalyptus*, with *E. diversicolor* (Karri), *E. globulus* (blue gum), *E. nitens*, *E. saligna* (Sydney blue gum), and *E. viminalis* (manna gum) particularly susceptible.

Distribution

The beetle is native to Australia but is also found in New Zealand, South America, South Africa and Zambia, the Middle East, the Mediterranean region, and in California, USA.

453 *Agrilus planipennis* (emerald ash borer) larval galleries under tree bark of *Fraxinus* sp. (ash). (Photo copyright of Jessica Metzger, Michigan Technological University, USA.)

454 Adult of *Phoracantha semipunctata* (eucalyptus longhorn borer) – note the buff-coloured patch on the wing case. (Photo copyright of C Fitzgerald, DPI Queensland, Australia.)

Impacts/ecology

The adult beetles are attracted to recently dead and dying trees or branches, where the females lay eggs under loose bark or in crevices. Young larvae either tunnel straight into the inner bark, or may feed at the surface, leaving a characteristic score mark, before moving under the bark. The attack is also characterized by sap bleeding. The damage results in wilting of foliage as translocation in the phloem is decreased or ceases, branch die back, and sometimes mortality results (455). The larval galleries are large (456) and may be up to 1 m long and 3 cm wide; so one gallery can ring bark and kill a tree (Hagen, 1999). Trees are particularly susceptible when under drought stress, but under good site conditions the beetle may not be a pest (Dreistadt *et al*., 1994).

Control

Cultural control, including the removal of cut logs and dead and dying material, can reduce the impact of the pest. The choice of good tree species for the site is the key to ensuring that stress and therefore susceptibility are minimized. Parasitoids from the beetle's native range, such as the Hymenoptera *Avetianella longoi* and *Syngaster lepidus*, show considerable promise in controlling populations across the current distribution (Hanks *et al*., 1995; Millar *et al*., 2002). For example, in California, 90% of eggs in the field have been found to be parasitized by the introduced *Avetianella longoi* (Hanks *et al*., 1995). Chemical control is regarded as unsuccessful for this pest, although *Bacillus thuringiensis* (*B.t.*), neem, and permethrin have been used on occasion.

ANOPLOPHORA GLABRIPENNIS MOTCHULSKY (COLEOPTERA: CERAMBYCIDAE: LAMIINAE: LAMIINI) – ASIAN LONGHORNED BEETLE
Description

The adults are large (20–35 mm long), shiny black beetles, with around 20 white dots on each wing case (457). The beetles bear the characteristic long antennae of the Cerambycidae (1.5–2.5 times body length); these antennae are striped with white bands. The legless larvae are cream in colour, up to 50 mm in length, and have a brown head capsule.

455 Stand of *Eucalyptus* sp. killed by *Phoracantha semipunctata* infestation in Queensland, Australia.

456 Galleries produced by *Phoracantha semipunctata* larvae in *Eucalyptus* sp. in Queensland, Australia.

457 Adult beetle of *Anoplophora glabripennis* (Asian longhorned beetle). (Photo copyright of James E. Appleby, Department of Natural Resources and Environmental Sciences, University of Illinois, USA.)

Host trees

In its native range, *A. glabripennis* is found on 23 species of *Populus* (poplars) and at least 24 other broadleaved trees, including some species of *Ulmus* (elms), *Salix* (willows), and *Acer* (maples/sycamores), as well as *Aesculus hippocastanum* (horse chestnut), *Malus pumila*, *M. alba* (apples), and *Prunus saliciana* (Yang *et al.*, 1995). In the northern USA, the beetle is known to attack *Acer* (maples) species in particular, although it has also been found on a range of genera, and is regarded as a specific threat to the maple sugar industry (USDA, 2002).

Distribution

The native range includes China, Taiwan, Korea, Japan, and parts of Southeast Asia. The beetle was first discovered in the USA in 1996 (Haack *et al.*, 1997a) and is present in New York City and Chicago. Although not yet confirmed as present in Europe, computer models predict that there is a significant risk of importation and establishment on the continent (MacLeod *et al.*, 2002).

Impacts/ecology

The adult beetles feed on leaves and twig bark, but do not cause significant damage. The larvae hatch from eggs laid under the bark and feed initially in the cambium during the second and third instar or larval stages (**458**). Infestations usually occur from the top of the tree downwards. As larvae mature they begin to tunnel into the sapwood and heartwood, where they pupate, with adults emerging from large round holes approx 1–1.5 cm in diameter (USDA, 2002). Consequently, the larvae not only inhibit nutrient and water flow within the tree, but also cause significant damage to the timber. The beetle is able to attack healthy as well as stressed trees, and death of the tree often results after three to five years (Smith *et al.*, 2005).

Control

Where infested trees are identified they are removed and destroyed; this is currently the main form of control. The efficacy of some systemic insecticides is being investigated, but none are currently used. There is potential to control the beetle using entomopathogenic (insect-infecting) nematodes such as *Steinernema feltiae* and *S. carpocapsae*, since these can detect and infect larvae that lie below the bark (Fallon *et al.*, 2004). There is also ongoing investigation of the potential of parasitoids, both those from the beetle's natural range and those that attack related beetle species in the areas invaded. The bethylid wasp *Scleroderma guani* and the colydiid beetle *Dastarcus longulus* may have potential as control agents (Yang and Smith, 2001; Smith *et al.*, 2005).

SHOOT BORERS

HYPSIPYLA GRANDELLA ZELLER (LEPIDOPTERA: PYRALIDAE) – MAHOGANY SHOOT BORER

Description

The adult moths have brown forewings and buff-coloured hind wings with a wingspan of approximately 2–4 cm. The forewings have a faint zigzag pattern and both forewings and hind wings have dark veins. Larvae are reddish-brown in colour with large black spots, and have a reddish head capsule with a black band behind (**459**). There are five to six instars (larval stages between moults).

458 Larval galleries under the tree bark excavated by the larvae of *Anoplophora glabripennis* (Asian longhorned beetle). (Photo copyright of James E. Appleby, Department of Natural Resources and Environmental Sciences, University of Illinois, USA.)

Host trees

These include a wide range of the family Meliaceae (sub-family Swietenioideae), including the commercially important *Swietenia* (American mahoganies), *Khaya* (African mahoganies), *Toona* (Burma cedar), and *Cedrela* (American cedars). The larvae feed in the shoots and fruits (Griffiths, 2001).

459 Larva of *Hypsipyla grandella* (mahogany shoot borer) on a host tree.

460 Damage to a stand of trees caused by infection with *Hypsipyla grandella*.

Distribution

H. grandella is the main pest of mahogany in the New World, central and tropical South America, the Caribbean, and Southern Florida, while *H. robusta* is the main pest in Africa and the Asia/Pacific region (Horak, 2001).

Impacts/ecology

The larvae tunnel in the developing shoots of young trees, frequently killing the leading shoot, which results in stunted, crooked growth (Griffiths, 2001); they may also feed under the bark. The larvae produce silk webbing in order to protect themselves from natural enemies and to improve their local microclimate. The generations last one to two months, extending to five months if the larvae go into diapause (a period of dormancy) triggered by cool temperatures or drought (Griffiths, 2001). Tree death can result from repeated attacks (460). Some adults feed on the fruit of the tree, causing premature fall.

Control

The impact of the shoot borer may be reduced by silvicultural methods, including pruning on high-quality sites to improve tree form, the production of trees which recover rapidly after attack, and the use of nurse crops (Hauxwell, 2001). The use of chemical control agents is generally restricted to integrated pest management (IPM) systems or to nurseries, as they are not effective on their own (Wylie, 2001). Parasitoids with the potential to control *H. grandella* have been identified, although further work on their biology is needed. Manipulation of native predatory ant populations, including *Oecophylla* spp. (weaver ants), *Anoplolepis* spp., and some *Iridomyrmex* spp., is thought to be worth exploration (Sands and Hauxwell, 2001).

TOMICUS PINIPERDA LINNAEUS (COLEOPTERA: SCOLYTIDAE) – COMMON PINE SHOOT BEETLE
Description

The adults are cylindrical and range from 3.5 to 4.8 mm in length. They have a black head and thorax, with reddish-brown to black elytra (wing cases). The anterior edge of the elytra is ridged or armed. The larvae are legless with a white body and brown head, and may be up to 5 mm in length.

Host trees

Various species of *Pinus* (pine) are the main host trees. When beetle populations are high, the adults may breed in *Picea* (spruces), *Abies* (firs), and *Larix* (larches) which occur in stands mixed with pine.

Distribution

The beetle occurs in Eurasia, North America (Illinois, Indiana, Maryland, Michigan, New York, Ohio, Pennsylvania, and West Virginia; and Ontario, Canada), Kunming, China, and Japan. It is found in sympatry (occurring in the same geographical region but without interbreeding) with *T. destruens* in France and the Mediterranean region (Haack *et al.*, 1997b).

461 Galleries created by *Tomicus piniperda* (common pine shoot beetle) under the bark of *Pinus nigra* (Corsican pine).

Impacts/ecology

There are two phases of attack. Initially adult beetles, which are associated with damaging blue stain fungi, will attack weakened trees and recently cut stumps and logs, using these as breeding sites. Volatile chemicals such as alpha-pinene are used to locate suitable material, and adults fly distances of up to 1 km to find it. Galleries are constructed for egg laying, and also by the larvae in the inner bark and outer sapwood (461). After pupation the adults emerge, leaving 2 mm diameter circular holes in the bark. The second and most economically significant phase of attack then occurs. Beetles fly to the crowns of healthy trees, where they carry out maturation feeding by boring into and hollowing out the centres of lateral shoots. The shoots bend, turn yellow/red, and break off (462, 463). Each adult can destroy up to six shoots, reducing productivity by up to 20–40% and affecting tree form (Fagerström *et al.*, 1978; Speight and Wainhouse, 1989; Haack *et al.*, 1997b).

Control

One of the most effective methods of control is to remove potential brood material of cut trees and stumps from the site. Trap logs may be set to attract adult beetles, and subsequently destroyed. Potential biological control agents include the predatory beetle *Thanasimus formicarius* (L.) (Cleridae) and species of *Rhizophagus* (Rhizophagidae) (Haack *et al.*, 1997b). A range of foliar chemical sprays including permethrin may be used to reduce beetle populations.

462 *Pinus sylvestris* (Scots pine) shoots attacked by *Tomicus piniperda* showing shoot bending and yellowing of their needles.

463 Numerous fallen pine shoots excised by *Tomicus piniperda* feeding on the young shoots (see 462).

BARK CHEWERS
HYLOBIUS ABIETIS LINNAEUS (COLEOPTERA: CURCULIONIDAE) – LARGE PINE WEEVIL
Description
The adult weevil is 9–14 mm long and dark brown, with cream-coloured markings on the elytra (wing cases) (**464**). The attachment of the antennae is at the tip of the snout, near the mouthparts (Byers, 1989). The adults live for up to four years and are nocturnal in habit (Heritage and Moore, 2001). A newly emerged adult is capable of reproduction within a month and eggs are laid underground, near a newly felled pine tree (Rose and Leather, 2003). The larvae are pale and legless with a brown head capsule (**465**).

Host trees
Many species of conifers, particularly *Picea sitchensis* (Sitka spruce), *Pinus sylvestris* (Scots pine), and *Larix* spp. (larches), act as hosts. The weevil is also a pest of several broadleaf tree species. Some trees such as *Pseudotsuga menziesii* (Douglas fir) and *Picea abies* (Norway spruce) are less vulnerable after two seasons' growth, whereas other species remain susceptible to attack (Rose and Leather, 2003).

Distribution
The pine weevil occurs in western and eastern Europe.

Impacts/ecology
This weevil is regarded as the most damaging and economically important pest of restocked forests in Europe (Rose and Leather, 2003). The larvae feed under the bark of stumps where they do no economic damage; this stage may last for two to five years. However, the adult *Hylobius abietis* is a significant pest, with two population peaks, spring and late summer, and therefore two periods of damage. In the UK, *Hylobius* is estimated to cost the Forestry Commission up to £2m per annum. Adults may live for four years and hibernate over winter, emerging in spring to feed. They feed on the bark and cambium in newly planted trees (**466**) – often at the base of the root collar in coniferous species and from the top of the stem in deciduous trees. Young trees are rapidly

464 Adult of *Hylobius abietis* (large pine weevil) – note the prominent antennae attached to the tip of the snout.

465 Legless larva of *Hylobius abietis*.

466 Bark damage caused by the feeding of *Hylobius abietis* adult.

ring barked and can be killed in a very short time (Heritage and Moore, 2001). The impact of this pest may be very serious, with weevils likely to kill up to 50% of newly planted trees on a site if no treatment is applied (Heritage and Moore, 2001).

Control

Cultural control is not effective for *H. abietis* as its populations increase rapidly and kill trees very quickly. Prophylactic chemical control is currently used in the form of permethrin sprays and dips (Rose and Leather, 2003). Biological control is being attempted using nematodes which locate and infect the host and rapidly develop inside it, resulting in septicaemia of the weevil caused by toxic bacteria (*Xenorhabdus*) accompanying the nematode. There are also some parasitoid wasps, mainly *Braconid* spp., which attack both the larvae and adults.

DEFOLIATORS

LYMANTRIA DISPAR LINNAEUS (LEPIDOPTERA: LYMANTRIIDAE) – GYPSY MOTH

Description

The adult female moth is large, white, and flightless, although winged. The wings have a dark zigzag pattern and a span of around 5 cm (**467**). The male has a wingspan of approximately 3–4 cm. The larvae have long hairs, and the final instars of the larvae are grey with five pairs of blue dots followed by six pairs of red dots along their backs (**468**; Humble and Stewart, 1994).

Host trees

The larvae feed on a very wide range of host plants (over 500 species). Young larvae feed primarily on broadleaf trees including *Quercus* (oaks), *Populus* (aspens/poplars), *Betula* (birches), *Salix* (willows), and *Alnus* (alders), whereas later instars can utilize a greater range of trees, including conifers.

Distribution

The gypsy moth is a native to Europe, southern Asia, and northern Africa. Its distribution includes British Colombia, Quebec, and Ontario in Canada, and Utah, California, and Washington State in the USA (Humble and Stewart, 1994).

Impacts/ecology

When populations are high, the gypsy moth defoliates millions of hectares of forest and urban trees. Outbreaks occur in the USA every seven to eight years, and can be three to four years in duration. The female moth lays a single egg mass on any available surface, including trees, rocks, fences, etc. The emerging larvae move to the tops of trees and are carried many miles on wind currents by ballooning on silk threads. Defoliation results in loss of productivity, and impacts on the recreational use of forests (**469**), but mortality is only likely if other stress factors for the tree are in play (Humble and Stewart, 1994). The caterpillars have histamine-carrying hairs, which can produce strong skin and respiratory reactions, so they are regarded as a major public health hazard.

467 Adult of *Lymantria dispar* (gypsy moth) showing the characteristic zigzag pattern on its wings. (Photo copyright of Thérèse Arcand, NRS Canada, CFS Laurentian Forestry Center.)

468 Larva of *Lymantria dispar* (gypsy moth) – note its long hairs, which can cause an acute allergic reaction in humans. (Photo copyright of Thérèse Arcand, NRS Canada, CFS Laurentian Forestry Center.)

Control

Mechanical methods of control can be used in urban areas, and include removing egg masses from trees and the application of grease bands to trees to prevent larvae moving down to the ground to pupate. A range of chemical control agents such as permethrin and diflubenzuron (a synthetic moulting agent) is effective in larger areas of outbreak. Biologically derived pesticides that are effective against gypsy moth (Sparks *et al.*, 1998) include the bacterial toxin *Bacillus thuringiensis* var. *kurstaki* (*B.t.k*), neem, and Spinosad (produced by the soil actinomycete *Saccharopolyspora spinosa*).

CHORISTONEURA FUMIFERANA CLEMENS (LEPIDOPTERA: TORTRICIDAE) – EASTERN SPRUCE BUDWORM

Description

The adult moths are greyish to copper-brown with a wingspan of approximately 2 cm (**470**; Cerezke, 1991). Females lay egg masses that are bright-green and found on the underside of needles. The young larvae are very small and light-yellow/green, becoming brown with a dark-brown head capsule (**471**). By the sixth instar they range in size from 1.8 to 2.4 cm in length. Pupation occurs in webbing in the tree foliage.

Host trees

This is a major forest pest and attacks firs and spruces such as *Picea glauca* (white spruce) and *Abies balsamea* (balsam fir), and, less commonly, *Larix* (larches) and *Pinus* (pines).

Distribution

The spruce budworm is native to North America and is found throughout most of Canada and the northern USA.

Impacts/ecology

Spruce budworm larvae damage trees by mining the current year's needles and new cones. Older larvae will occasionally feed on mature needles. The female moths lay eggs in masses on the underside of needles in August, and the peak feeding period for larvae is April to June. Feeding results in defoliation, giving the trees a reddish-brown tinge. The leading stem may

469 Defoliation damage caused by *Lymantria dispar* (gypsy moth) to a stand of broadleaved trees. (Photo copyright of Claude Monnier.)

470 Adult of *Choristoneura fumiferana* (eastern spruce budworm). (Photo copyright of Thérèse Arcand, NRS Canada, CFS Laurentian Forestry Center.)

471 Larva of *Choristoneura fumiferana* (eastern spruce budworm). (Photo copyright of Thérèse Arcand, NRS Canada, CFS Laurentian Forestry Center.)

be killed, affecting tree shape. Seed production may be reduced and productivity is lowered (**472**). Where trees are susceptible (due to age or condition), mortality may occur after several years of defoliation (Cerezke, 1991).

Control
Cultural practices may help to reduce impact, and include planting even-aged, low-density stands, maintaining vigorous growth, and early harvesting (Unger, 1995). In Canada, tebufenozide (Mimic®) is used as an aerial chemical control spray (Cerezke, 1991; Unger, 1995). Pheromone (attractant chemical) traps are used to monitor population and dispersal trends of adult moths, together with aerial surveys of defoliation. *B.t.k* is used as a biological control agent and is optimally applied when the larvae are 10–15 mm long, usually in June.

MALACOSOMA DISSTRIA HÜBNER (LEPIDOPTERA: LASIOCAMPIDAE) – FOREST TENT CATERPILLAR

Description
The adults are pale yellow to reddish-brown with a wingspan of approximately 2.5–4.5cm. The forewings are marked with two diagonal lines, which may form a dark band (**473**). The larvae are dark brown and hairy with blue to blue-black sides, reaching 4.5–5.5 cm by the last instar. The dorsal surface is marked with whitish, keyhole-shaped spots bordered by two orange lines. There is a brightish blue band along the side of the larva (**474**; Wood, 1992).

Host trees
The larvae feed on deciduous trees such as species of *Populus* (poplars/aspens), *Salix* (willows), *Alnus* (alders), *Betula* (birches), *Prunus* (plums), and *Quercus* (oaks).

Distribution
The forest tent caterpillar is native to and widespread in North America.

472 Damage to a young tree shoot caused by *Choristoneura fumiferana* (eastern spruce budworm). (Photo copyright of Claude Monnier.)

473 Adult of *Malacosoma disstria* (forest tent caterpillar). (Photo copyright of Thérèse Arcand, NRS Canada, CFS Laurentian Forestry Center.)

474 Larva of *Malacosoma disstria* (forest tent caterpillar). (Photo copyright of Thérèse Arcand, NRS Canada, CFS Laurentian Forestry Center.)

Impacts/ecology

In spite of the name, the larvae do not build a tent, but create a mat on which they feed in groups. Outbreaks of tent caterpillar occur about every 10 years and can last anything from three to seven years. The larvae begin to feed at budburst and may kill the buds, causing trees to grow with poor form and to lose twigs and branches. Feeding by older larvae may result in complete defoliation, reducing tree productivity and occasionally causing mortality (**475**). However, death is only likely if a tree succumbs to drought, other pests, or disease, or after several subsequent years of defoliation (Wood, 1992).

Control

Egg masses and larvae are often cut out of trees in urban areas. In forest situations infestations may be left to take their course as populations are eventually reduced by adverse environmental conditions (e.g. frost), starvation, and natural enemies. However, *B.t.k* and a range of chemical pesticides, including permethrin, may be used to control the caterpillars. An NPV (nuclear polyhedrosis virus) is a key natural agent of mortality in the wild, and is currently being explored for its potential as a control agent (Frank and Foltz, 1997). Over 80% of the tent caterpillar pupae may be parasitized and killed by *Sarcophaga aldrichi* Parker (tachinid flesh fly), but Hymenopteran parasitoids are unlikely to control populations. However, bird predation, for example that by *Icterus galbula* (northern oriole), has a significant impact on densities (Parry *et al.*, 1997).

TERMITES

Termites are social insects, living in colonies, with several castes or body forms specialized to perform specific tasks, such as soldiers, workers (absent in primitive families), and reproductively active, winged (alate) forms. Termites live in nests which are constructed differently across groups; in the genus *Coptotermes*, nests are built underground (in living/dead trees or in man-made structures) from a combination of soil, chewed wood, saliva, and faecal material (collectively called carton). Soil is often compacted around the outside.

COPTOTERMES FORMOSANUS SHIRAKI (ISOPTERA: RHINOTERMITIDAE) – FORMOSAN SUBTERRANEAN TERMITE AND *COPTOTERMES ACINACIFORMIS* FROGATT (ISOPTERA: RHINOTERMITIDAE)

Description

Soldier termites are often used for identification of *Coptotermes* species because they are proportionately numerous in the colonies. *Coptotermes formosanus* soldiers have oval orange-brown heads and black jaws or mandibles. The body of the soldier is creamy-white and up to 6.5 mm in length (**476**). The winged forms (alates) are yellow-brown, large in size at 12–15 mm long,

475 Defoliation damage caused by *Malacosoma disstria* (forest tent caterpillar) to a stand of broadleaved trees. (Photo copyright of Pierre Therrien.)

476 Termite soldiers of *Coptotermes formosanus* showing their large mandibles (Photo copyright of Nan-Yao Su, Fort Lauderdale Research & Education Center, University of Florida, USA.)

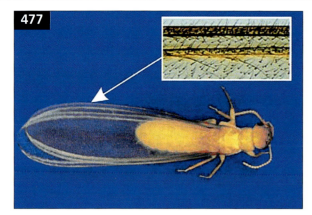

477 Alate (winged reproductive) termite of *Coptotermes formosanus* showing hairs on its wings. (Photo copyright of Nan-Yao Su, Fort Lauderdale Research & Education Center, University of Florida, USA.)

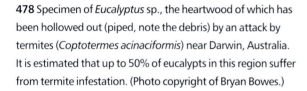

478 Specimen of *Eucalyptus* sp., the heartwood of which has been hollowed out (piped, note the debris) by an attack by termites (*Coptotermes acinaciformis*) near Darwin, Australia. It is estimated that up to 50% of eucalypts in this region suffer from termite infestation. (Photo copyright of Bryan Bowes.)

and have many small hairs across the wing surface (477; Scheffrahn and Su, 1994). *Coptotermes acinaciformis* (tree piping termite) is of a similar size to *C. formosanus* (5–6.6 mm long), and has a teardrop-shaped head and tapering toothless mandibles.

Host plants

C. formosanus attacks a wide range of plant species, living and dead, as well as man-made constructions and non-cellulose-based materials; it is often found in damp locations (Cabrera *et al.*, 2001). *C. acinaciformis* attacks and hollows out (or pipes) live trees, and *Eucalyptus* species are known to be particularly susceptible (478). It also attacks human constructions.

Distribution

C. formosanus is thought to be endemic to southern China and is currently found in Japan and Hawaii (Tamashiro and Su, 1987), South Africa, and Sri Lanka. In the USA it has been found in Alabama, California, Florida, Georgia, Hawaii, Louisiana, Mississippi, North and South Carolina, Tennessee, and Texas. (http://creatures.ifas.ufl.edu/urban/termites/formosan_termite.htm).

C. acinaciformis is native to Australia, and has been found in isolated colonies in New Zealand, but to date these colonies have been successfully eliminated.

Impacts/ecology

The large size of *Coptotermes formosanus* colonies (there may be as many as several million individuals), makes this species a highly destructive pest. Most infestations of man-made structures result from the termites colonizing upwards from their underground nests; however they are also able to construct aerial nests when conditions are appropriate (Su and Scheffrahn, 1987). Infestation is often quite advanced before signs such as soft floorboards and blistered tree surfaces are apparent. *C. formosanus* is regarded as one of the most significant pests in the USA (http://creatures.ifas.ufl.edu/urban/termites/formosan_termite.htm).

C. acinaciformis is the most economically important termite in Australia, causing significant damage to various species of *Eucalyptus*. It is also a major pest of urban forestry and human constructions. *C. acinaciformis* can build subsidiary nests, connected by underground galleries or tunnels to the main nest, and has an extensive foraging range.

Control

Aggregation and control – one of the main methods of controlling termites – involves encouraging them to aggregate at a food source or 'bait', and then eliminating them using chemicals such as arsenic trioxide or metabolic inhibitors such as sulfuramid (Peters and Fitzgerald, 2005). Physical barriers and chemical barriers, such as a soil drench of pyrethroid chemicals, may be effective, although chemicals are short-lived and application must be repeated. The use of more resistant building timber such as *Callitris* (cypress pines) can be useful (Peters *et al.*, 2005). The role of entomopathogenic fungi in the control of *Coptotermes formosanus* is being investigated. Fungi may be of use in an IPM system, when combined with bait systems (Wright *et al.*, 2004).

CONCLUDING REMARKS

Insect herbivores are part of a complex ecological system, in which they utilize plant material and are themselves exploited by predators and parasites. A great many groups of insects are herbivorous – over 50% of extant species feed on plants (Speight *et al.*,

1999) and therefore have the potential to be regarded as insect pests. The list of insect pests is rising steadily. Given that a greater area of global forest cover is being exploited every year and that there is increased movement of people, materials, and insects around the globe, it is likely that the list will continue to grow. The future for insect pest control lies in an IPM strategy that takes into account the system in which the insect operates, clear assessment of actual damage to the forest, and a range of effective and appropriate monitoring and control methods, including cultural, biological, and chemical methods.

SECTION 5 FOREST ECOLOGY AND MANAGEMENT, TREE SURGERY, PROPAGATION, AND CONSERVATION

CHAPTER 10

General forest ecological processes

Peter A Thomas

INTRODUCTION

Wooded land currently covers between 30% and 35% of the world's land surface (depending on what is counted as forest), or around 39–45 million km² (FAO, 2003). Ecologists often distinguish between woodland and forest (**479, 480**). Woodland is a small area of trees with an open canopy (usually defined as the canopy giving less than 40% cover, that is 60% or more of the sky is visible) so that plenty of light reaches the ground, encouraging other vegetation beneath the trees. By contrast, a forest is usually considered to be a relatively large area of trees forming a closed, dense canopy. For simplicity's sake, and because the underlying ecological processes at work are the same, in this chapter the term forest will be used to mean any wooded land.

SIZE AND GROWTH

The most obvious factor that separates forests from other types of habitat is the large weight or mass of organic material present, referred to as the biomass (or sometimes the standing crop). In most forests, more than 85% of the biomass is contained in the above-ground portion of the woody plants. Biomass above ground increases from the northern boreal forest southwards towards the tropics, starting from very low levels at the Arctic tree line, and reaching in

479 A dense temperate rain forest in the Olympic Peninsula, Washington State, USA. It is composed mostly of *Pseudotsuga menziesii* (Douglas fir), *Tsuga heterophylla* (western hemlock), and *Thuja plicata* (western red cedar).

excess of 940 t ha^{-1} in the Amazon basin. However, there are exceptionally large forests outside the tropics, notably the temperate forests of the Pacific Northwest of North America. These include stands of huge *Psuedotsuga menziesii* (Douglas fir, **479**), reaching 1,600 t ha^{-1}, and *Sequoia sempervirens* (coastal redwoods), the tallest trees in the world, which have a biomass of up to 3,450 t ha^{-1} just in the trunks. Below-ground biomass in roots is significantly less (Jackson *et al.*, 1996), averaging 29 t ha^{-1} in boreal forests, 40–42 t ha^{-1} in temperate and tropical deciduous forests, and 49 t ha^{-1} in tropical evergreen forests.

Biomass is a static measure of how much mass there is at any one time, with no indication of how quickly new growth is being added or lost, and so gives little insight into how the forest is functioning. More useful are estimates of the productivity of the forest, i.e. how much new material is being added per year, described as net primary productivity (NPP). This can vary from as little as 1 t ha^{-1} y^{-1} in cold boreal forests, to over 30 t ha^{-1} y^{-1} in tropical rainforests, with an average of 7–12 t ha^{-1} y^{-1} in temperate forests. However, a maximum of 36.2 t ha^{-1} y^{-1} has been recorded in the Pacific Northwest from a 26-year-old forest of *Tsuga heterophylla* (western hemlock). These figures have sometimes been used to calculate how much additional forest needs to be planted to soak up (sequester) the huge amount of extra carbon that is being pumped into the atmosphere – usually approximately 25% extra forest globally. However, such an estimate is blatantly wrong. When a forest is mature it reaches an approximately steady state of mass, where NPP is balanced by an equal loss in biomass through decomposition. At this point, the productivity of the whole forest (the net ecosystem productivity – NEP) drops to near zero. Thus, it is only young forests that are carbon sinks; once forests are mature they become carbon neutral. In reality, temperate and northern forests globally are a net sink of carbon, but this is primarily due to expansion of the amount of forest due to reforestation (Beedlow *et al.*, 2004).

LIGHT

Trees have evolved as a life form to outcompete their neighbours for light by growing tall, so producing dense forests that inside are darker, more humid, and less prone to extremes of temperature variation than outside. In temperate forests at least, it is usually possible to recognize four reasonably distinct layers (**480**). At the top is the tree canopy, normally 5+ m above ground. Below are the shrub layer (<5 m), the field or herb layer of herbaceous plants and short woody plants such as brambles, and the ground or moss layer of mosses and liverworts, lichens, and algae. Each layer blocks sunlight so that a dense layer may

480 Open woodland at Needwood Forest, England. The sparse canopy of *Fraxinus excelsior* (ash) and *Tilia* spp. (lime) standards allows abundant light to reach the shrub layer of coppiced *Corylus avellana* (hazel) and, on the ground, a mixed field layer dominated by *Hyacinthoides non-scripta* (bluebell). A sparse ground layer of mosses is also present.

preclude any layers below, and the forest floor may be very dark indeed (**481, 482**). In temperate regions, the amount of light reaching the forest floor may be as high as 20–50% of full sunlight in an open birch wood, down to just 2–5% beneath *Fagus sylvatica* (European beech). In these deciduous forests, light levels are higher once the leaves have fallen, but the trunks and branches still block some light such that light levels are likely to be below 70–80% of full sun. Evergreen forests tend to cast similar shade all year round; in Europe, light levels below natural *Pinus sylvestris* (Scots pine) forests are usually around 11–13%, while below *Picea abies* (Norway spruce) they can be as low as 2–3%. In tropical rain forests, light levels at the forest floor may be even lower, just 0.2–2% of full sunlight.

As a rule of thumb, plants require 20% of full sunlight for maximum photosynthesis and at least 2–3% sunlight for photosynthesis to exceed background respiratory costs (the compensation point). This inevitably means that the floor of densest forests is at, or beneath, the limits of plant growth. Some forest floor plant specialists have overcome this problem with a number of physiological solutions.

- Using shade leaves that are thinner and more efficient at low light levels than sun leaves.
- Reducing the compensation point. Bates and Roeser (1928) found that coastal redwood in deep shade requires just 0.62% sunlight.
- Making use of sunflecks – patches of sunlight passing through gaps in the canopy – which can briefly give up to 50% of full sunlight and make up 70–80% of the total solar energy reaching the ground in a dense forest (Evans, 1956). These flecks are especially important to shade plants that are capable of responding quickly to the brief flurries of light.

Plants can also cope with dark conditions by avoidance. Temperate deciduous forests are well-known for their colourful carpets of prevernal plants, which grow and flower early in spring. In the UK these include *Hyacinthoides non-scripta* (bluebell), *Ranunculus ficaria* (lesser celandine), and *Anemone nemorosa* (wood anemone, **483**). These plants make use of the light reaching the ground

481 Woodland of *Quercus robur* (pedunculate oak) with a very well-developed ground layer of mosses and a sparse field layer of ferns beneath the oak canopy. The shrub layer is missing, due to heavy sheep grazing. Dartmoor, England.

482 Canopy of *Taxus baccata* (yew), which is so dense that no other layers can grow beneath it. Box Hill, England.

483 *Anemone nemorosa* (wood anemone), a prevernal plant of deciduous woodlands growing in the UK.

before the trees develop their canopy of leaves, and die back once the shade is too deep. Summergreen plants, such as *Mercurialis perennis* (dog's mercury) and *Galium odoratum* (woodruff), are similar but keep their leaves through the summer using what little light is available. As an extension of this strategy, wintergreen plants (which keep at least a few green leaves all year round) and true evergreen plants can start growth as soon as spring conditions allow, and continue growth into a warm late autumn after leaf fall. Such plants include wintergreen *Oxalis acetosella* (wood sorrel) and *Primula vulgaris* (primrose, 484), and evergreens such as *Hedera helix* (ivy) and *Ilex aquifolium* (holly). Being evergreen is an efficient strategy for coping with seasonally abundant light, but it does carry costs. In winter, holly is a sitting target for herbivores such as deer, and so has evolved prickly spines to the leaves. These spines are absent above deer-browsing height, around 3 m above ground.

Tree seedlings face similar problems of shade, having to grow up through dark layers of vegetation before reaching the canopy. Different tree species vary tremendously in how much shade they can bear as seedlings and saplings. *Fagus sylvatica* (European beech, 485) and *Acer saccharum* (sugar maple, from North America) are very tolerant of deep shade, while *Betula* spp. (birches) and *Populus* spp. (poplars) grow best under high light intensities. However, it is now apparent that the ability to tolerate shade can change through the lifespan of a tree (Poorter *et al.*, 2005), so it is possible that many trees are more shade-tolerant as seedlings than as adults.

Nevertheless, comparatively few trees can tolerate the full shade cast by their mature relatives. Consequently, they depend upon gaps appearing in the forest, by one or more trees dying or falling (486), for successful establishment of seedlings. Gaps are sufficiently important that while large-scale regional vegetation (e.g. oak forest) is determined by climate, soil, and topography, it is the dynamics of gaps that largely controls the proportions in which the various species grow in any one area. For example, in small gaps created by one tree falling, shade-tolerant trees such as *Fagus* spp. (beech) or *Abies* spp. (fir) are more likely to do best and dominate. In larger gaps, species

484 *Primula vulgaris* (primrose), a wintergreen plant that keeps some leaves alive throughout the year.

485 Seedlings of *Fagus sylvatica* (European beech) are very shade-tolerant and capable of growing under the dense canopy of their parents. Each seedling shows two distinctively shaped cotyledons below the young shoot.

486 *Pinus nigra* var. *maritima* (Corsican pines) uprooted by the wind, producing a characteristic pit and mound topography, with the exposed mineral soil ideal for seedling establishment. Delamere Forest, England.

such as *Betula* (birch) and *Salix* (willow), which invade quickly from light, wind-borne seeds and grow rapidly, are more likely to dominate initially but later give way to shade-tolerant trees. It is not just what goes on above ground that is important; in larger gaps there will also be less below-ground competition from the root systems of the large trees at the gap edge. The importance of such competition has been demonstrated experimentally by cutting roots (trenching) around the edges of a plot: seedlings inside the plot usually grow faster (e.g. Barberis and Tanner, 2005). Competition may also happen below ground from the field layer vegetation by allelopathy, i.e. secretion of chemicals, which inhibit other root growth, into the soil (e.g. Orr *et al.*, 2005). Further variability in seedling establishment is produced by small-scale heterogeneity of the forest floor. Pits and mounds of bare mineral soil created by falling trees (486) offer less competition and a more constant water supply than the surrounding humus-rich forest floor. In a *Pinus sylvestris* (Scots pine) forest in Finland, Kuuluvainen and Juntunen (1998) found that although these bare sites covered just 8.4% of the forest, they held 60% of pine and 91% of birch seedlings and saplings. Dense field and ground layers can cause problems for tree regeneration, swamping small seedlings. This is one reason why, in temperate rainforests, seedlings are often most common on 'nurse logs', which are continuously damp enough to provide moisture and lift the seedlings above the dense field layer (487, 488).

As tree seedlings grow upwards into a gap, there can be intense competition to reach and keep the light; whichever seedlings grow quickest will dominate the gap, at least in the short term. A common strategy to get a head start, found in trees as diverse as *Fraxinus excelsior* (European ash), and shade-tolerant firs (Narukawa and Yamamoto, 2001), is to have a seedling bank. Here, young plants survive in light conditions below their compensation point (i.e. they are sustaining a net loss of energy) and grow very slowly while their energy reserves last. These seedlings are then able to take rapid advantage of an opening in the canopy in the race for dominance.

487, 488 A sapling of *Tsuga heterophylla* (western hemlock) growing on top of a nurse log in the temperate rain forest of the Olympic Peninsula, Washington State, USA (**487**). The rotting log provides abundant moisture and freedom from competition from the field layer. The sapling has now rooted into the ground and is independent of the nurse. Years later (**488**) the nurse log has rotted away, leaving the new tree and other western hemlocks in a curiously straight line.

WATER

Given that a single, large deciduous tree can use 400,000 litres of water in transpiration in a summer (Thomas, 2000), it is obvious that whole forests move immense amounts of water from the soil to the atmosphere. Nevertheless, water is rarely limiting for tree growth in temperate regions until rainfall decreases to such an extent that scrub and grasslands take over. Almost all roots tend to be quite shallow, so potential problems exist if the surface layers of the soil are drained of available water between rain events. This is obviated, however, by the process of hydraulic lifting present in a number of trees and a few grasses. Here, water is raised at night from moist areas lower in the soil (flowing along a hydraulic gradient through the roots) to nearer the surface. Hydraulic lifting is most common in savannas and other xeric (dry) woodlands, especially among older trees (Domec *et al.*, 2004), but is found elsewhere. The amounts moved can be significant: a mature *Acer saccharum* (sugar maple) 19 m high can raise around 100 litres of water each night compared to a water loss via transpiration of 400–475 litres the following day (Emerman and Dawson, 1996). This raised water also benefits other surrounding plants (Penuelas and Filella, 2003; Filella and Penuelas, 2003–2004).

Forests also play a significant role in the redistribution of water on a regional scale. Rainfall intercepted by the canopy is evaporated before it reaches the ground. When this and the transpiration of water are combined (evapotranspiration), the overall losses are in the order of 30–60% of precipitation in deciduous forests, 50–60% in tropical evergreen forests, and 60–70% in coniferous forests, compared to around 20% in grasslands. Not surprisingly, forested areas have water yields (measured as stream flow) 25–80% lower than pastures. Moreover, computer modelling by Calder *et al.* (2003) suggests that planting oak woodland in central England would eventually reduce recharge of aquifers and runoff to streams by almost one half. So, should forest be removed to improve water yield? Most data show that regardless of forest type, removal of up to 20% of the trees has an insignificant effect on water yield, presumably because of increased soil evaporation replacing evapotranspiration (Brown *et al.*, 2005). Further clearance does improve water yield (Bosch and Hewlett, 1982), but by comparatively small amounts until clearance is significant.

Many people have held the view that forests increase rainfall in a watershed through evaporating water, thus helping build clouds. However, in temperate areas, at least, the contribution of a forest to rainfall is likely to be insignificant and certainly less than 5% (Golding, 1970). On a continental scale, forests help to increase rainfall in the sense that they repeatedly recycle the atmospheric moisture passing from the oceans to the land. For example, in the Amazon Basin, much of the daily rainfall is immediately evaporated to generate clouds for rainfall downwind. It is highly likely that continual clearance of the forest will reduce rainfall elsewhere in the region since much of the water will enter rivers and be lost to the system. Moreover, the effects of such tropical deforestation have far wider repercussions in mid- and high latitudes through large-scale links in the water cycle and weather. Avissar and Werth (2005) have shown, for example, that deforestation of Amazonia and Central Africa severely reduces rainfall in the Midwest of the USA.

NUTRIENTS

Nitrogen is usually the nutrient most limiting growth in temperate forests, while in other forests, especially on soils of great age, phosphorus may well be the limiting nutrient. Nutrients within a forest ecosystem are highly recycled and key to this recycling are the decomposer organisms that release nutrients from dead material. Larger soil fauna, such as earthworms and beetles, chew debris into fine particles suitable for the soil fungi and bacteria. A square metre of soil in temperate woodland may contain more than 1,000 species of animal, from protozoa to earthworms, and a gram of soil can contain more than 1,000 species and more than 200 million bacterial cells (Fitter, 2005).

Soil organic matter (surface litter and humus incorporated into the soil) is thus the main bottleneck controlling nutrient availability to plants, and the slower decomposition is, the more of a limiting factor it is. This helps explain why slow plant growth occurs on cold northern soils that have large organic matter accumulations.

Fungi and bacteria are not altruistic in providing nutrients to plants. As dead material is decomposed, nutrients released by the micro-organisms are immediately taken back up by other micro-organisms, and so are effectively immobilized and unavailable to plants. However, as the carbon is progressively used up in their respiration (and released as carbon dioxide), the conserved nutrients become more than the microbes can use, and the excess is released in inorganic form for plants to use. Consequently, when a fresh batch of litter arrives on the forest floor there is a variable time lag before its carbon has been reduced sufficiently to allow nutrients to be freed into the soil for plant growth, the process being regulated by the microbial community (Attiwill and Adams, 1993; Ågren et al., 2001). Plants can, however, circumvent this bottleneck in several ways. Firstly, more than 80% of the world's vascular plants have on their roots mycorrhizal fungi, which greatly assist in scavenging nutrients from the soil to the symbiotic benefit of both plants and fungi. Secondly, some plants are now known to be able to directly use organic nutrients, without the intervention of micro-organisms first breaking them down into inorganic forms. For example, up to 50% of the total nitrogen in forest soils is usually in the form of dissolved organic nitrogen (DON), of which approximately 10–20% consists of amino acids. The degree to which plants can use DON is open to speculation, but it is becoming clear that many plants are capable of absorbing amino acids directly (Lipson and Näsholm, 2001) and are thus able to short-circuit the micro-organism bottleneck. The same may also be true for organic phosphorus.

Although nutrients are tightly recycled within a forest ecosystem, there are still (usually small) annual inputs and losses. Nutrients are added to forests through rain and dust, dissolved from rocks in the soil, and as biological input from nitrogen fixation by microbes. Losses of nutrients can be very rapid due to fire, wind, and erosion but the majority of losses, from temperate forests at least, are by leaching of nutrients as water percolates through the soil. However, since nutrients are vital to forest growth, plants and microbes are fairly efficient at reabsorbing and holding available nutrients and creating conditions of controlled decomposition. This has been admirably demonstrated by the Hubbard Brook

489 Watershed No. 2 of the Hubbard Brook Ecosystem Study in the White Mountain National Forest of New Hampshire, USA. The hardwood forest of the watershed was clear-felled in 1965–1966 (and treated with herbicides for three years to prevent any regrowth) to investigate how water flow and nutrient loss in the drainage stream changed. (Photo copyright of USDA Forest Service, Northeastern Research Station.)

Ecosystem Study in the White Mountain National Forest of New Hampshire, established in 1963 (**489**; Likens, 2004). As part of this, a discrete watershed was clear-felled in 1965–1966 and treated with herbicides for three years to prevent any regrowth, while a similar watershed had the hardwood forest left intact. After clear-felling, stream flow went up (due to reduced evapotranspiration) and net losses of nitrate, calcium, and potassium in stream water generally peaked in the second year, each returning to pre-cutting levels at rates unique to each ion as the forest regrew. However, even decades after clear-felling, differences in stream water solutes can still be seen, especially in calcium (Likens et al., 1998).

There is still a good deal to learn about mechanisms of nutrient retention in forests. For example, Muller and Bormann put forward the vernal dam hypothesis in 1976. This proposes that prevernal plants, which grow early in spring before canopy closure, take up nitrogen and other nutrients before they can be leached; these are subsequently made available to other plants as the prevernal plants die back from lack of light. At Hubbard Brook, plants of *Erythronium americanum* (yellow trout lily) saved almost half of the important nutrients from being washed away. In the spring they used 43% and 48% of the released potassium and nitrogen, respectively, with the rest being lost in stream water.

490 One of the nitrogen enrichment plots in a stand of *Pinus resinosa* (red pine) at Harvard Forest, Massachusetts, USA. The various markers and flags show where repeat samples of soil and litter are taken.

Some subsequent experiments (e.g. Tessier and Raynal, 2003) have supported the theory. However, other contradictory studies have shown that the microbe population itself is better at soaking up the spring burst of nutrients (e.g. Zak *et al.*, 1990). Also, while the dying back of vernal plants can produce a burst of nutrients (e.g. Anderson and Eickmeier, 2000), the plants may not be very efficient at taking up nutrients in the first place (e.g. Anderson and Eickmeier, 1998; Rothstein, 2000). Undoubtedly, some of the experimental differences come from investigating different plant species in several forests.

The tight recycling of nutrients within the forest ecosystem can cause problems if too much arrives as pollution. Nitrogen enrichment, particularly in northern temperate areas, is just such a case (Nosengo, 2003). Since the 1980s, normal background nitrogen deposition of <1 kg ha^{-1} y^{-1} has increased by 10–40 times or even higher. The effect of too much nitrogen is clearly seen in long-term experiments running at Harvard Forest, Massachusetts since 1988 (Magill *et al.*, 2004). In one of these, a plantation of *Pinus resinosa* (red pine, **490**) was subjected to three levels of nitrogen (**491–493**): a control, low N addition, and high N addition. After 14 years, annual wood production had decreased by 31% and 54% relative to the control in the low N and high N plots, respectively, and the canopies had thinned due to dieback under higher nitrogen levels. Mortality also increased (control 12%; low N 23%; high N 56%) and the whole high N stand was expected to die in the near future.

491–493 The effect of 14 years of nitrogen enrichment on *Pinus resinosa* (red pine) at Harvard Forest. **491**: control plot with no extra nitrogen added above the background deposition of 7–8 kg N ha^{-1} y^{-1}; **492**: low N addition (50 kg N ha^{-1} y^{-1}); and **493**: high N addition (150 kg N ha^{-1} y^{-1}).

COARSE WOODY DEBRIS

The vital importance of dead wood in forest carbon budgets, and also as an invaluable wildlife resource, has been increasingly appreciated over the last decade (Kirby and Drake, 1993). Dead wood appears in many forms, sizes, and positions including standing dead trees (snags), dead branches in the canopy, and trunks and branches on the ground. A useful term for this motley collection is coarse woody debris (CWD). Typically, CWD in a forest forms up to a quarter of all the above-ground biomass and is normally in the range of 11–38 t ha^{-1} in deciduous forests, with the largest amounts in cooler regions where decomposition is slower. Conifer forests generally hold more CWD than deciduous forests, typically around 100 t ha^{-1}, but up to 500 t ha^{-1} in the coastal redwood forests of California and the rain forests of the Pacific Northwest (**494**). Tropical forests, with more rapid decomposition, usually have lower amounts of woody accumulation, but levels up to 100 t ha^{-1} are possible in more water-logged areas of the Amazonian forest. If 100 t ha^{-1} of wood was spread evenly over the forest floor it would amount to 10 kg in each square metre. However, because the bulk of the wood is in large pieces, typically less than 5% of the ground will be covered by CWD, although this can rise to around a third cover in very dense coniferous forests. Snags are of particular wildlife interest. In the Białowieża forest of Poland, one of the most pristine forests in Europe, Bobiec (2002) found that standing dead wood varied from 3 to 21% of total CWD (**495**), and figures of 25% are typical in many of the world's forests.

Wood is difficult to decompose. It is composed of 40–55% cellulose, 25–40% hemicelluloses, and 18–35% lignin (conifers having a greater proportion of lignin than hardwoods). Wood is thus high in structural carbohydrates (which require specialized enzymes to break them up) but also poor in nutrients such as nitrogen: 0.03–0.1% N (by mass) compared to 1–5% in foliage.

In most forests, wood (CWD) will be colonized by fungi within a year and completely colonized within 5–10 years. However, decay rates of wood vary tremendously depending upon the climate, decaying organisms available, and the size and type of wood. In general terms, pioneer trees such as birches and willows invest less energy in protecting their wood from rot (going for speed of growth rather than defence) and logs on the ground rot away within a few decades. Wood from longer-lived trees such as oaks may persist for a century or much longer, while in cool climates such as the Pacific Northwest (**494**) wood may persist for up to 600 years (Franklin *et al.*, 1981). Even in tropical rain forests, wood above 3 cm diameter takes at least 15 years to decompose (Anderson and Swift, 1983). Again, however, environmental conditions play an important role in

494 Temperate rain forests, such as the one here on western Vancouver Island, Canada, can contain large quantities of dead wood, in part because of the size of some of the fallen logs. The one shown here is of *Picea sitchensis* (Sitka spruce).

495 The Białowieża forest in Poland during winter shows the large amount of dead wood that can build up in a forest, from small twigs through to fallen trees and broken dead snags.

determining decay rates; logs of *Populus balsamifera* (balsam poplar) in North America, which would decay away within 40–60 years on land, last for over 250 years when water-logged in a beaver pond.

EVERGREEN AND DECIDUOUS LEAVES

At first sight, the occurrence of evergreen and deciduous trees in different forests can appear haphazard, but in reality it demonstrates the interactions of many of the ecological processes described above (Thomas, 2000). Deciduous trees lose their leaves during an unfavourable season (winter in temperate areas), while evergreen trees always have some leaves on the tree and individual leaves may live from six months to over 30 years. If growing conditions are favourable all year round, as in tropical rain forests (**496**), then there is no selective advantage in being deciduous and so evergreen angiosperms dominate. In climates with a dry summer or cold winter, it is cheaper to grow thin disposable leaves than to grow more robust leaves capable of surviving the off-season, so in most moist temperate areas deciduous trees dominate (**497**). However, if environmental conditions become worse, it may once again be more beneficial to grow evergreen leaves. This includes areas with a very short growing season, where evergreen leaves are able to start growing as soon as conditions allow and

so none of the growing season is wasted. This accounts for evergreen leaves in northern and alpine areas (**498**), and also among woodland understorey

496 Evergreen foliage is typical of areas where the climate is hospitable for growth all year round, such as in this tropical rain forest in Malaysia.

497 Deciduous forest in Harvard Forest, Massachusetts, USA. In a seasonal temperate climate it is more economical for trees to grow a set of disposable leaves each spring rather than build leaves capable of surviving the winter.

498 Evergreen conifers, such as *Abies lasiocarpa* (subalpine fir) shown here in the Canadian Rocky Mountains, are typical of areas with short growing seasons where deciduous trees are disadvantaged by wasting part of the season producing new leaves.

shrubs such as holly and ivy, which benefit from an early spring start and late autumn finish when the canopy has no leaves. Evergreen leaves are also found in Mediterranean climates (**499**) where the winter growing season is dry; leaves that are protected enough to cope with the droughty conditions will also survive the hot dry summer, and so effectively become evergreen and need to be kept for several years to repay the high investment cost. In areas where the climate becomes even more severe, such as at the Arctic treeline or in alpine areas, deciduous leaves re-appear. Despite the problems of a very short growing season and acute shortage of nutrients, the winter is so severe that it is cheaper to build new leaves every year rather than attempting to keep leaves alive. Thus, the northernmost trees in the Arctic and uppermost trees in alpine areas are deciduous trees such as species of *Betula* (birch), *Larix* (larch, **500**), and *Salix* (willow, **501**).

499 Evergreen species are also found in Mediterranean climates, such as the coastal foothills of California shown here, where the growing season is hot and dry and trees require tough expensive leaves to survive. The main species shown are *Pinus sabiniana* (gray, or digger, pine) on the right, and the darker green *Quercus douglasii* (evergreen blue oak).

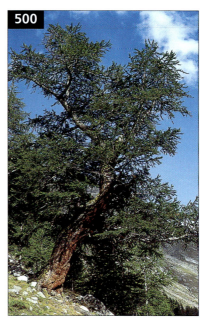

500 *Larix decidua* (European larch), growing here in the Swiss Alps, is deciduous and one of the last tree species to be met before reaching the treeless alpine areas above.

501 In the very short growing season of the tundra overlying permafrost, evergreen shrubs give way to deciduous *Salix* spp. (willows). (The graves are those of 19th century whalers who overwintered and died here on Herschel Island in the Arctic Ocean.)

CONCLUDING REMARKS

Forest ecosystems work in much the same way as any other ecosystem, but size and complexity create ecological situations that are unique to forests. The large amounts of biomass that can be grown in a year appear useful for carbon sequestration in relation to global warming, but must be weighed against the decompositional losses in mature forests, and possibly the extra methane – a potent greenhouse gas – that these will generate (Keppler *et al.*, 2006). To maintain sequestration rates, new forests are constantly needed. Light availability presents problems for those plants living below the dense forest canopy, but these problems are solved by making do with less light or growing when light is available in the spring or during brief sunflecks. The role of forests in the water cycle still needs to be fully clarified, but it is of great importance due to the likely pressure on forests as human water needs increase. Nutrient dynamics in forests are crucial to their long-term well-being and it is important that we improve our understanding of the effects of climate change and pollution on decomposition and nutrient cycling. Of necessity, this chapter gives only a résumé of a very large subject. A more detailed account of forest ecology is provided by Thomas and Packham (2007).

CHAPTER 11
Silvicultural systems

Peter Savill and Nick Brown

INTRODUCTION

A silvicultural system is a planned series of treatments for tending, harvesting, and re-establishing a stand. The main systems, their variations, and applications are described in this chapter. Forests in different parts of the world have been managed in a huge variety of ways to achieve different mixes of products and benefits, using locally appropriate methods. While no single system is ideal for all situations, there are no fundamental differences in the principles of silviculture when applied to tropical or temperate forests, plantations, or pristine natural forest. Silvicultural systems are often flexible and imprecisely defined but, in general terms, there is a continuum of possibilities from the creation of exotic plantation forests through to low-impact manipulation of natural forests. The flexibility of silviculture is necessary to take account of changes in the demand for products and services that will occur during the time it takes for a stand to grow. Flexibility may also be required to accommodate changes in the environment, legislative framework, political priorities, and technology.

ECOLOGICAL PRINCIPLES IN FOREST REGENERATION

The assemblage of trees in a forest changes continually as individuals grow, die or are killed, and are then replaced by others. This turnover results in a patchwork of stages of succession, which reflects the past sequence of tree deaths. Forests subject to large disturbances that kill many trees at the same time, such as forest fires or storms, tend to be composed of large patches of even-aged trees. In contrast, in forests with a low magnitude and frequency of natural disturbance, many trees will die of old age. This creates gaps the size of a single tree, and results in a very heterogeneous forest.

The larger a gap, the more the microclimate differs from that of the surrounding forest. The size of the gap is one of the main determinants of which tree species will eventually fill it. Large gaps are colonized by species that are ecologically adapted to exploit open, disturbed conditions. Small gaps tend to be filled slowly by shade-tolerant species that may have established in deep shade.

Silvicultural systems are typically designed to optimize conditions for timber production and/or the development of desirable forest services (such as biodiversity conservation). When trees are cut from a forest, gaps in the canopy are created (**502**). The size of the gaps will depend upon how many trees were cut, and the care with which they were removed. Patterns of harvesting should be designed to create conditions that match the establishment requirements of the next generation of trees. When single trees are felled they create

502 A small felling gap in lowland dipterocarp rain forest, Sabah, Malaysia. A small gap such as this will promote growth of shade-tolerant seedlings but avoid triggering an invasion of light-demanding pioneers and climbers.

503 Emergent trees, 50–60 m tall, soar above the canopy of lowland rain forest in Sabah, Malaysia.

small gaps in which it is likely that a shade-tolerant species will succeed. Some smaller tree species never grow to the height of the canopy and spend all their lives in the shade, but these are commercially unimportant. However, some heavy hardwood species (densities >880 kg m^{-3} at 15% moisture content), such as *Neobalanocarpus heimii* (chengal), a rain forest species from Malaysia, may eventually grow to be massive (approximately 50–60 m tall) emergents containing large volumes of high-quality timber (503). In larger gaps, species that are initially more light-demanding, or are able to respond rapidly to higher light levels by fast growth, are likely to out-compete more tolerant or slower responding tree species. These are often the light hardwood species (densities 400 to 720 kg m^{-3} at 15% moisture content) such as *Swietenia macrophylla* (mahogany), which can have high commercial values when present in sufficient quantities. A large clear-felled gap will generally favour 'pioneer species', most of which have very light wood and seldom reach very large dimensions. Many such trees are of low commercial value, but those that grow to larger sizes and have more valuable wood are commonly used as plantation trees in clear-cutting systems (e.g. most pines, *Terminalia ivorensis*, and *Gmelina arborea*).

In general, species adapted to regenerate in the crowded conditions under canopies or in small gaps do not make successful species for the clear-cutting system. They can be difficult to establish, and tend to be poorly formed and to grow much more slowly than colonizers, and so require more protection from browsing and more weeding. However, once established they can be very productive. Such species are more effectively managed by silvicultural systems that minimize clear-felling – such as some shelterwood and selection systems. They are also valuable for underplanting older crops to obtain forests with two distinct age classes.

It is not surprising, therefore, that the ecological basis of many silvicultural systems lies in the manipulation of the forest canopy. By controlling canopy gap size, it is possible, to some extent, to determine the species composition of the next growth cycle.

Table 11 A key to the major silvicultural systems

1	a	Crop trees are of one or two age classes and all mature trees are harvested in a single felling, or over a few years. Regeneration occurs across the whole stand at the same time.	Even-aged systems – see 2
	b	Crop trees of a wide range of ages are grown together. Trees are felled when they reach maturity and regeneration occurs in felling gaps, maintaining an uneven-aged (irregular) structure.	Selection (or uneven-aged) systems – see 4
2	a	Stands regenerated by planting or from seed, seedlings, or advance regeneration.	High forest systems – see 3
	b	Stands regenerating from stool shoots or suckers.	Coppice systems
3	a	Mature crop trees are removed in a series of fellings over relatively few years to provide sheltered growing space for regeneration while maintaining some seed trees. This gives rise to an approximately two-aged stand for a period in the regeneration cycle.	Shelterwood systems
	b	All crop trees are harvested in a single felling and regeneration occurs simultaneously across the entire stand, giving rise to an even-aged forest.	Clear-cutting system
4	a	Individual mature trees felled leaving surrounding forest untouched.	Single tree selection systems
	b	Groups or strips of forest felled.	Group or strip selection systems

CLASSIFICATION OF SYSTEMS

Silvicultural systems differ in four major ways (*Table 11*). The most important difference is whether the crop trees fall into one (or a few) distinct age classes, or have a wide range of ages. Clear-cutting and coppice systems, where all stems are harvested simultaneously over a large area, are extreme even-aged systems. At the other extreme, in the single tree selection system only single trees are extracted, leaving the surrounding forest virtually intact. Only the largest cohort of trees is harvested, leaving smaller sized individuals to grow on to fill the space. Felling and regeneration occur continuously throughout the stand and this maintains or creates an uneven age structure.

The second important difference is in the method of forest regeneration used. This can be vegetative (from coppice or root suckers), or by planting seedlings, direct seeding, or natural regeneration. Natural regeneration can be from newly established seedlings, or pole-sized trees (advance regeneration).

Although coppice systems are clearly distinct, most other silvicultural systems can use any, or a combination, of the other techniques.

The third difference between systems depends on whether all crop trees are removed in a single felling operation before a new crop is established, or whether a proportion of the mature trees are retained to provide seed and shelter for regeneration.

The final difference is whether a treatment is applied uniformly to the whole stand, or whether groups or strips of trees are managed at different times.

EVEN-AGED SYSTEMS
COPPICE SYSTEM

Coppice shoots arise primarily from concealed dormant buds that grow from the stump of a tree following cutting (**504**). They can also develop from buds on roots in some species to give rise to root suckers, and a few reproduce by both methods (e.g. most acacias, *Balanites aegyptica*, and *Daniellia oliveri* in dry parts of Africa).

The coppice system relies upon these methods of vegetative reproduction after each stand of trees has been felled to provide the next generation. Coppice regeneration has an advantage over seedlings in that ample supplies of carbohydrates are available from the parent stool and its root system, so new shoots grow very vigorously from the start. However, coppice shoots of most species seldom grow to the dimensions of trees grown from seed, so the system is used primarily to produce small-sized material. The ability to coppice is far more common in broadleaved trees than in conifers. Species also vary greatly in their vigour of coppicing: poplars, willows, and eucalypts are generally very good at regenerating. The longevity of a stool varies with its health, species, and site. Some are relatively short-lived, lasting only two or three rotations (for example, *Eucalyptus*), while others (such as *Tilia*) are almost indestructible. Among suitable species, no method of regeneration has a greater certainty of such rapid and complete success, and in the rather rare circumstances today where coppicing is profitable, no other method of regeneration is cheaper. The system can be attractive financially because coppice rotations are much shorter than those in high forest where trees are grown from seed.

Variants of coppicing include coppice-with-standards, pollarding, and shredding, the latter two being mostly associated with isolated trees rather than woodland.

• Woodlands managed as coppice-with-standards usually consist of simple, even-aged coppice as the underwood, and an overwood of standards, which are normally trees of seedling rather than coppice origin (505). The standards are uneven-aged and the two components of the system have quite different rotation lengths, so providing both large and small stems from the same piece of land. Coppice-with-standards is the oldest of all deliberately adopted systems of forest treatment in most temperate parts of the world. Cuttings are made in both the overwood and the underwood at the same time. When the coppice underwood has reached the end of its rotation and is cleared, standards which have reached the end of their productive life are also removed and new ones introduced.

504 Coppice shoots growing from willow (*Salix* sp.) stumps in Oxford, England. Willow of this kind is increasingly being grown as a short-rotation crop for electricity generation.

505 Oak coppice with standards in Germany, with the coppice due to be cut for fuelwood, after growing for about 25 years. The larger, single-stemmed trees are 'standards', of seedling origin. They occur in a range of ages (sizes) that correspond roughly with intervals between coppicing.

• Pollarding involves cutting trees 1.5 to 3.5 m above the ground, rather than at ground level, and allowing them to grow again (**506**). This puts the regrowth out of reach of cattle and other browsing animals. Today, pollarding is often done for garden/landscape ornament. Any tree species that can be coppiced will respond to pollarding, except those where root suckers are depended upon for regeneration.

• Shredding involves the repeated removal of side branches on a short cycle, leaving just a tuft at the top of the tree. It was formerly practised in Europe, on land where there was little grass, to feed cattle on the harvested leafy shoots, and especially with elm. Today it is sometimes carried out in countries with Mediterranean or monsoon climates, such as parts of Nepal, where there is a long, dry, grassless season (**507**). The deeper rooting shredded trees can provide ample fodder from their leaves.

Coppicing is one of the oldest forms of forest management, but has been in decline in many temperate regions since at least the mid-1800s as a result of industrialization. Plastic, metal, and other alternatives now replace the many objects and implements formerly made from wood of small dimensions. In the developed world, improvements in infrastructure for distributing gas, electricity, oil, and coal also mean that wood is seldom required as a fuel. In its modern form, coppice is normally worked on a clear-cutting system and is extensively used for the production of pulpwood (e.g. from *Eucalyptus*), and for short-rotation energy crops (from *Salix* and *Populus*), as well as for fuelwood. The latter usage is particularly important in the tropics where, for example, *Combretum* spp., *Acacia seyal*, and *Leucaena leucocephala* are cultivated (**508**).

506 An old pollarded willow tree (*Salix fragilis* L.) in Oxford, England. Pollarding is carried out partly for ornament to maintain the traditional landscape, and partly to prevent browsing by cattle. The regrowth occurs above the reach of cattle.

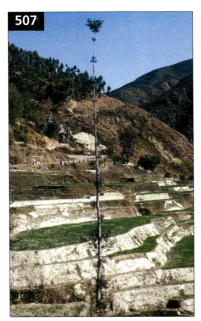

507 A 'shredded' tree in Nepal. Shredding involves the repeated removal of side branches on a short cycle, the leafy shoots providing animal fodder during the long, dry, grassless season.

508 Coppicing of *Leucaena leucocephala* (Lam.) de Wit for production of fuelwood. In this example, the leaves were also used for feeding goats. Ibadan, Nigeria.

SHELTERWOOD SYSTEMS

The essential feature of the shelterwood system is that even-aged stands are established, normally by natural regeneration, under a thinned overstorey that produces sufficient shade and a moderated environment for young trees to establish themselves. The overstorey is removed as soon as establishment is complete. The success of a uniform system depends upon establishing high stocking levels of new seedlings of commercially desirable species before the final felling occurs.

Progressive opening up of the forest canopy provides space and light for regeneration but does not expose seedlings to drastic changes in microclimate nor to competition from faster growing, more light-demanding weed species. Regeneration for the next rotation arises almost entirely from seedlings and, among some pioneer species, from seed in the forest floor (e.g. the normally unwanted *Musanga cecropioides* in West Africa). Damage to the forest is much more drastic than in selection systems: the canopy is extensively removed, and bigger gaps are formed, favouring light-demanding species. As this type of system aims to nurture light-demanding species (such as many of the dipterocarp species found in southeast Asian rain forests), it does not depend on advance regeneration from pre-existing saplings and poles of commercially important species in the understorey. Little advance regeneration of these species will be found beneath a closed forest canopy anyway, and those saplings that have managed to survive there are often damaged during harvesting. Advance regeneration of young trees is usually removed at the time of the seeding felling (see below).

Treatments in this system usually include the following.

- Preparatory felling: a late thinning to encourage the development of the crowns of future seed bearers.
- Seeding felling: once it is clear that there is going to be a good seed crop, a third to a half of the trees are removed together with the understorey and any regeneration already present. Cultivation may be carried out to assist seedling establishment (**509**).
- Secondary fellings: there are usually two to four fellings carried out at three to five year intervals, but with their timing and intensity carefully regulated to allow seedlings to grow while also preventing vigorous weed growth (**510**).
- Final secondary felling: when the remaining overstorey is removed.

509 Shelterwood system with oak (*Quercus robur* L.) in the Loire Valley, France. A seeding felling has just been carried out, removing a third to a half of the trees. The understorey and any regeneration already present have also been removed, and the ground has been cultivated to assist seedling establishment.

510 Shelterwood system with oak (*Quercus petraea* Mattuschka Liebl.). A late secondary felling allows prolific natural regeneration growth. Bellême Forest, Normandy, France.

The damage to regeneration in later fellings is not usually serious, especially if the regenerating trees are young, supple, dense, and even-aged (**511, 512**). The whole series of operations normally takes 5–20 years, but infrequent mast years or frost-sensitive seedlings both necessitate longer regeneration periods. For a light-demanding species, the secondary fellings must be few and rapid, and the whole process may be completed in five years. If seed production is infrequent, then it may take up to 50 years to obtain adequate regeneration. The stand will then be much more uneven-aged and patchily distributed, in which case the system grades into the group shelterwood (see below).

One of the main advantages of the shelterwood system is said to be its simplicity, but in areas where mast years are infrequent, obtaining a fully stocked, even-aged regeneration is a major managerial problem. The shelterwood system can also be used with planted stock where natural seeding is insufficient or irregular, where a change of species is required, or where seed-bearers are insufficient in number or quality. Stands managed under a shelterwood system have many features in common with those established by planting under a clear-cutting system. They can be pure, even-aged, and uniform in structure and density over large areas.

GROUP SHELTERWOOD SYSTEMS

Group shelterwood systems involve a similar retention of an overstorey for a short period to provide shelter for the new stand, which is approximately even-aged (**513, 514**). The main difference from a uniform shelterwood, apart from the smaller sizes of the areas worked, is that if advance or existing regeneration is present, it is used as the focus of a regeneration felling. Groups are gradually enlarged by carrying out regeneration fellings (seeding, secondary, and final fellings successively) around the edges until eventually they meet and merge. The regeneration period is generally longer (15–40 years) than with the shelterwood system, and the resulting stand is therefore somewhat more uneven-aged. Another variant is the uniform strip system, which consists of shelterwood regeneration fellings carried out in a strip ahead of the advancing edge of the final felling; this is sometimes considered most appropriate for light-demanding species.

CLEAR-CUTTING SYSTEM

In this system almost all trees in a stand are felled in a single operation. Often, strips of uncut forest are left along the banks of rivers for erosion control and along public roads for aesthetic reasons. Clear-cutting is the predominant silvicultural system in

511 *Pinus roxburghii* Sargent being managed under a uniform shelterwood system at Naini Tal, Uttar Pradesh, India in 1912. Before felling, the mature, seed-bearing trees were tapped to death for resin.

512 Shelterwood system with oak (*Quercus robur* L.) just after the final felling, showing a large area of dense, even-aged regeneration. Vrbovec Forest, Croatia.

plantations managed primarily for the production of large timber volumes and in boreal forests, where it emulates natural large-sized disturbances. Its main advantages include simplicity, uniformity and, in particular, ease of felling and extraction. The system almost always operates with establishment by planting, which has the advantage of its artificiality and minimum reliance on the unpredictable quality and species of the regenerating trees. However, clear-cutting does not necessarily preclude the use of natural regeneration, as in the seed tree system variant of clear-felling (where a small number of widely spaced adult trees are retained for seed, **515**) and strip-clear-felling (**516**).

513 A group shelterwood system in Kelheim Forest, Bavaria. The main species are Norway spruce (*Picea abies* L. Karsten), European silver fir (*Abies alba* Miller), beech (*Fagus sylvatica* L.), Scots pine (*Pinus sylvestris* L.) and European larch (*Larix decidua* Miller). Use has been made of advance regeneration, which was the focus of the characteristically dome-shaped group, with the tallest regeneration in the middle. This picture was taken in 1913, but the system remains almost unchanged today.

514 A group shelterwood system near Munich, Bavaria. The group on the right of the picture is now very large, having been increased in area by carrying out successive regeneration fellings (seeding, secondary, and final fellings) around the edges. Eventually groups meet and merge. The regeneration period is about 40 years and the resulting stand is consequently somewhat uneven-aged. The main species are Norway spruce (*Picea abies* L. Karsten) and European silver fir (*Abies alba* Miller).

515 Western larch (*Larix occidentalis* Nutt.) seed trees retained after clear-cutting a lodgepole pine stand near Cranbrook, British Columbia, Canada. (Photo copyright of Roger Whitehead, Natural Resources Canada, Canadian Forest Service.)

516 Strip-clear-felling in *Pinus sylvestris* L. This is a clear-cutting system that relies on natural regeneration rather than planting. The width of cleared strips is about four times the height of the seed-bearing trees, which can be seen in lines. France, 1955.

517 Clear-cutting system: a poplar plantation near Isparta, Turkey. Intensive production of clonal poplars represents one of the most artificial forms of forestry. The ground vegetation is virtually eliminated by cultivation, and the trees are irrigated during dry periods of the year. As a consequence, production is extraordinarily high, at around 40 m³ ha⁻¹ y⁻¹.

major motive, clear-cutting is invariably chosen (**517**), unless some social, biological, or environmental factor of the locality rules it out.

GROUP CLEAR-CUTTING

This involves felling all the trees in a group prior to restocking. The stand within each group will always be even-aged, but the stand as a whole will contain groups of a wide range of ages, and possibly of all ages. The individual groups may be pure or mixed in species composition, and may be established by natural regeneration, planting, or a combination of both. Group clear-cutting is particularly appropriate to strong light-demanders, as the only protection given to the young trees is from side shelter. Group sizes commonly range from about 50 to 110 m in diameter (0.2–1.0 ha).

UNEVEN-AGED SYSTEMS
SINGLE TREE SELECTION SYSTEMS

The tropical 'polycyclic' system is based on the repeated removal of selected trees in a continuing series of felling cycles, the length of which is less than the time it takes the trees to mature (Whitmore, 1990). It is analogous to the selection system of temperate forests. The very species-rich nature of many tropical rain forests, and the relatively small number of species with timber that is commercial by current standards, means that the polycyclic system tends to result in the formation of scattered small gaps in the forest canopy. Its success depends upon the following.

- A good supply of half-grown shade-tolerant trees at the time of felling.
- Removal of nearby mature trees of commercial value, while ensuring that enough are left to produce seed.
- Half-grown trees must not be seriously damaged by harvesting.

Generally, the bigger the trees and the larger the number that are harvested, the greater the disturbance or damage caused to the residual stand and the greater the chance that light-demanders will form a significant part of the future stand, at the expense of the more shade-tolerant species.

The polycyclic system involves the manipulation of a forest to maintain a continuous cover, and to provide for the regeneration of the desired species, and the controlled growth and development of trees

However, planting is expensive, losses may be high (especially through drought), and stocking is usually orders of magnitude lower than with good natural regeneration. The latter may result in >500,000 young trees ha⁻¹, whereas plantations seldom have more than 5,000 young trees ha⁻¹. Hence, the resulting planted stand may be of lower quality. The disadvantages of clear-cutting (rather than of planting) largely arise from the lack of protection to the site. This may lead to soil erosion, a rise in the water table, extremes of temperature including frost, leaching with soil acidification, and rank weed growth. Clear-cutting is widely regarded as the least desirable system for both landscape and conservation but these disadvantages can be reduced by the use of small group fellings (0.2–2 ha).

Clear-cutting is based strongly upon principles of economics and finance. It provides good opportunities for using labour-saving equipment and machinery efficiently; management is simple, and work can be carried out with little skilled supervision. Management can, in fact, be intensive, and hence cost-effective. For production systems where profit is a through a range of diameter classes, which are mixed singly (in the single tree selection system) or in groups (group system). Successful management can be very complex. It depends on a sound ecological knowledge, experience (in which considerable intuition may be involved), and silvicultural judgement. It aims for the maintenance of a stable and apparently relatively unchanging forest environment.

Stands managed on a polycyclic system are, at all times, a mixture of trees of all age classes. There is no concept of a rotation length, or of a regeneration period, as both harvesting and rc-cstablishment take place regularly and simultaneously throughout the stand. The only silvicultural interventions are 'selection fellings', which are typically carried out every 5–20 or more years throughout the stand. These fellings are a combination of regeneration tending, cleaning, thinning, final felling, and regeneration felling. This can be difficult, as the needs of each of the age classes must be taken into account and trees of all sizes are removed. An important feature of selection felling is that it concentrates on improving the quality of the stand, rather than felling to remove the largest and best stems, which may result in impoverishment. Many humid tropical forests have been harvested by 'high-grading', where only the largest and most valuable trees are removed with little regard to stand improvement. The residual forest is often highly disturbed, of little productive use, and vulnerable to conversion to other types of land use.

Without careful intervention, there is usually a tendency for a more even-aged structure to evolve, and also for the different age classes to become spatially separated, so that a group structure develops. In an extreme case, this would result in even-aged, single-storeyed groups. This occurs with light-demanding species, and such a 'group selection' system is the only form of the selection system appropriate to them.

The length of the period between successive selection fellings varies. Short periods (less than five years) allow good stand management, particularly of young trees. Damage to the canopy in terms of excessive opening is also reduced, although damage to the soil by heavy machinery is likely to be increased. Long periods result in larger volumes of timber being removed at each visit, making them more financially economical. They also improve the success of regeneration of light-demanders because the canopy is opened up more.

518 Single tree selection forest at Couvet, Switzerland. The main species are Norway spruce (*Picea abies* L. Karsten) and European silver fir (*Abies alba* Miller).

Polycyclic systems are appropriate for the management of tropical moist and semi-deciduous forests, where the majority of species are shade-tolerant and few species have commercial value. Harvesting is extensive, with usually only a small number of stems felled per hectare. The best continental European examples are in the silver fir (*Abies alba*) forests (with *Fagus* and *Picea abies*) of central Europe (**518**). In temperate regions, selection systems are largely confined to mountainous areas, where a continuous protection of the soil against erosion (and often against avalanches) is of great importance. Selection systems also protect the soil against leaching and are suitable for regeneration of frost-sensitive tree species. Selection forests are probably the ideal for conserving landscapes, and appropriate for forests around towns, where an apparently unchanging view is important. However, contrary to popular belief, they do not necessarily even approximate to 'natural' forests in many places where they are applied.

GROUP STRIP, WEDGE, AND EDGE SYSTEMS

The term group selection is widely used and loosely applied to any irregular or group system. It should strictly refer only to systems in which a stand is sub-divided into groups, each of which is, for a large part of its life, uneven-aged, and has more than one storey. Group selection systems are also referred to as irregular shelterwood systems (Matthews, 1989).

220

In all group systems, the size of the group is a critical characteristic. Large groups are easier to manage, and are essential for light-demanders. The most useful range is probably 0.1–0.5 ha, larger groups being needed in taller and more uneven-aged stands. The shape and orientation of the groups can have a considerable influence on the variation of microclimate within them, and considerable emphasis is laid on this in central Europe. General observations are that a north–south orientation of an elliptical or rectangular group provides a good compromise between wind and sun, and that light-demanders should be near the north edge (in the northern hemisphere), and frost-tender tree species near the south.

The layout of groups is vital in facilitating management of the stands. Wherever possible, the first groups to be regenerated are those located furthest from the road, thereby minimizing the amount of timber that has to be extracted through a young stand. Fencing costs for small groups are inordinately high and this has always been considered a major disadvantage of any group system.

Group systems come closest to imitating the structure of a natural stand, at least in many temperate regions, and are therefore increasingly recommended for use.

The various group systems can all be applied as variants of the three main high forest systems (shelterwood, clear-cutting, and selection), giving group clear-cutting, group shelterwood and group selection systems. A whole compartment of a group clear-cutting system may therefore be uneven-aged, but each individual group will be even-aged and managed on a clear-cutting system. Similarly, strip systems can be considered as variants of each of the three basic high forest systems, depending on the type of stand treatment that is carried out ahead of the advancing felling edge. This gives, for example, strip-clear-cutting (**516**), and strip-shelterwood systems. Other shapes, including wedges, very occasionally replace strips, and these are described by Matthews (1989).

SILVICULTURAL SYSTEMS IN DIFFERENT PARTS OF THE TROPICS

While the principles of silvicultural systems remain the same everywhere, the various constraints assume different levels of importance in different types of forest (*Table 12*).

Table 12 Constraints to successful regeneration in different types of forest

Type of forest	Main constraints
Tropical moist forest	• Inadequate stocking of seedlings and saplings of desirable species and poor seed production • Excessive damage to natural regeneration caused by careless and unplanned timber extraction • Invasions and subsequent poor control of climbers and scramblers (e.g. *Merremia* spp. in Borneo)
Tropical dry forest	• Seasonal, but very uncertain, precipitation, long dry periods, heat, risk of fire • Predominantly range areas for cattle, goats, camels, etc., so browsing a major problem • Soils often alkaline
Mountain forest	• Possibilities of severe soil erosion after felling • Steep terrain – difficult access, harsh climate, slow growth, storms, avalanches, rock falls, browsing
Temperate broadleaved forest	• Poor seed production, prolific weed growth, deer browsing
Temperate coniferous and boreal forest	• Harsh climate, slow growth, often acid, organic soils, storms, browsing

TROPICAL MOIST FOREST MANAGEMENT

Different forest areas are subjected naturally to different degrees of gap creation. It is believed, for example, that in much of Amazonia large gaps are seldom created by natural events, and consequently most regeneration is of shade-tolerant sub-canopy species. The occasional shade-tolerant heavy hardwood becomes emergent. In parts of southeast Asia and tropical Australia, where cyclones are relatively regular, the creation of large gaps is more common. Here, a far higher proportion of the natural climax forest is composed of light-demanding species. The forests of West Africa are said to be intermediate. Hence, rain forests are not uniform the world over, and consequently different silvicultural systems have been found to be appropriate to different parts of the humid tropics.

A characteristic of tropical rain forests is that they grow slowly. Growth rates of the commercially saleable tree species in West African forests, for example, seldom exceed $1.0–1.5 \, \text{m}^3 \, \text{ha}^{-1} \, \text{y}^{-1}$. If all trees in the forests are considered, the yield would increase to about $4 \, \text{m}^3 \, \text{ha}^{-1} \, \text{y}^{-1}$. Commercial growth rates are therefore little different from those in the coniferous boreal forests of Sweden or Canada, and maximum yields do not exceed those found in temperate broadleaved forests.

Silviculture can be successful so long as it is practised within the biological limits of the forest. Regrettably, the concept of a sustained yield does not necessarily coincide with an economist's view of maximum profitability. Accountants require high volumes to be removed at one time to make the operation profitable, preferably the entire merchantable crop from every area worked. This may not replicate the processes of gap formation and regeneration natural to the area, and can result in considerable damage to the forest (**519**). For counties with impoverished economies, it appears more 'profitable' – or politically important – to adopt systems based on short felling cycles and large volumes removed per unit area. This usually results in serious over-cutting, and sometimes complete loss of the forest; in short, the forests are mined, rather than managed sustainably.

519 Widespread damage to lowland dipterocarp rain forest in Borneo caused by careless high-lead logging. Very few seedlings of commercially valuable species were left undamaged and regeneration was mostly of light-demanding pioneer tree species and climbers.

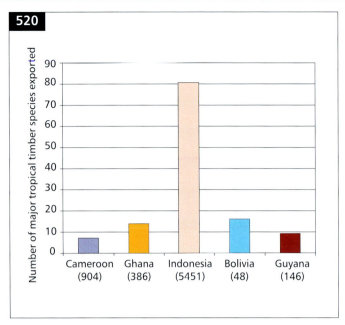

520 Number of major timber species exported (as logs, sawn timber, veneer, and plywood) from five tropical countries. Figures in parentheses are the volumes of tropical timber exported (1000 m³) in 2003. Data from ITTO (2004). The rain forest in southeast Asia has much higher stocking of commercially valuable timber trees than other tropical regions, and many more species are marketable.

Southeast Asia

In Malaysia, the Philippines, and parts of Indonesia, the dominant tree family is the Dipterocarpaceae. The rate of growth and form of dipterocarp species mean that they produce high-quality timber ranging from light to heavy hardwood. The wood of many dipterocarp species can be grouped into a small number of timber types (i.e. meranti, red, and seraya, yellow). This means that a much larger number of species can be marketed than is the case in West Africa or the Neotropics. As a consequence, the volumes of timber extracted per unit area of a southeast Asian rain forest can be two or three times greater than in the former regions (**520**).

Many dipterocarps produce seed gregariously, and in great quantities, at intervals of 3–11 years. The seedlings survive low levels of light, but unless a canopy gap forms overhead, most will die within a few years. The implication of this is that it is crucial that the forest is already well stocked with seedlings of desirable species when the forest is logged, and that logging does not destroy these seedlings. If a logged forest lacks seedlings of desirable species, then regeneration will be composed primarily of non-commercial species or colonists.

Before the Second World War there was a steady demand for naturally durable construction timber from the heavy hardwood dipterocarps. This favoured management on a polycyclic system, with very few stems being removed per unit area, and resulted in good regeneration of shade-tolerant dipterocarps. After the war, the demand for timber increased significantly and, at the same time, species with lighter wood were required. In Peninsular Malaysia these formed a large proportion of the trees in the lowland forests. A monocyclic system was established, termed the Malayan Uniform System (MUS). This involved a pre-felling inventory to establish that there were sufficient seedlings of desirable species, felling of all marketable trees in a single operation, and a post-felling girdling of unwanted species to open the canopy (Wyatt-Smith, 1963; in Mergan and Vincent, 1987). Extensive canopy opening encouraged the regeneration of fast-growing (approximately 60–80 years rotation), light-demanding species for which there was a large and lucrative market. This system converted large areas of the tropical lowland rain forest into a more or less even-aged forest containing a high proportion of commercial species. The system depended upon the following.

- An initially adequate and well distributed stocking of seedlings.
- More or less complete removal of the original canopy.
- No further tending until after the ephemeral climber stage had passed.

- Treatments to ensure maintenance of an adequate newly regenerated canopy.
- Regular forest sampling to assess the status of the regeneration.

In the mid-1970s, economic pressures arising from shortages of farmland led to much of the lowland forest being cleared. The MUS was largely replaced by a polycyclic system in the hill forests. The MUS never fitted well to these forests because of their difficult terrain, uneven tree stocking, and lack of natural regeneration after harvesting. In addition, heavy seedling mortality occurred due to logging on steep slopes, and the seeds of commercial species had poor viability. The selective management system (SMS) was eventually developed to deal with these problems, especially in forests that were still well stocked with adolescent trees. SMS is a polycyclic system with a felling cycle of 25–30 years. It is flexible and claims to be based on the forest characteristics. However, the selection procedure involves the optimization of management goals, which tend to have been hijacked by economists, rather than controlled by ecologists and silviculturists. Wyatt-Smith (1987) considered that SMS would produce an acceptable yield of timber in its second cut, but thereafter yields would decline due to the lack of large-diameter trees. Many foresters consider that polycyclic systems will generally be unable to maintain timber yields over a number of felling cycles, due to the level of damage to the residual forest, and the relatively short periods usually allowed for the forest to recover.

The Philippines and Indonesia introduced polycyclic systems, with the amount of timber removed at any one cycle being controlled by tree diameter. These remain in use today, but the forests have suffered damage from poor enforcement of the diameter limits, and breached restrictions on road building. In northern Queensland (Australia), a polycyclic system was developed in which tree felling approximately every 40 years was to be practised, combined with strict control of road construction, felling size, and direction of felling. Although it appeared to be a model system, it is no longer in operation because the region where it was developed has been declared a World Heritage Area, in which all exploitation is banned.

West Africa

A detailed account of the history of silviculture in West Africa has been written by Parren (1991). These forests have a longer history of management than in the neotropics, but shorter than those in southeast Asia. In general, they do not contain the high proportion of valuable tree species (520) that is found in Malaysia, but are better provided than in the forests of the neotropics. Light-demanding species occur in some abundance in natural forests.

Monocyclic systems, originally transferred from Malaysia and developed in Nigeria in the 1940s, were known as the tropical shelterwood system (TSS) and modified selection system. Management was both labour- and cost-intensive, and involved cutting back climbers, suppressing non-commercial tree species, and promoting the natural regeneration of commercial species by a judicious mix of cutting and poisoning. The aim was that in the five years prior to exploitation, at least 100, one metre high, seedlings should be established per hectare. The forest could then be logged (removing 65–80% of the basal area), and the 100 seedlings would be tended with cleaning and thinning operations over the next 15 years.

The systems were largely abandoned by the mid-1960s because of inability to cope with climber growth, and the recognition that at least some poisoned species subsequently became very commercially valuable. Also, the growth of the released seedlings was not adequate to justify the costs involved in tending them. Regrettably, this abandonment has often resulted in over-exploitation or complete loss of the forest to shifting cultivation.

The Ghana polycyclic selection system, dating from 1956 and lasting to 1971, involved stock mapping of all economic trees with diameters over 67 cm at 1.3 m from the ground (or above any taller root buttresses). Younger trees were thinned, and selective fellings took place over a 25-year cycle. The system was eventually abandoned due to the slow growth of commercially important trees (with trunks averaging growth of 0.6 cm diameter y^{-1}), high costs, and the increasing scarcity of valuable species following selective logging. The current challenge for Ghana is to shift logging to new, more available tree species, and also to increase the rotation length for the more valuable timbers such as *Chlorophora excelsa* (iroko) and *Khaya* spp. (African mahogany) so that they can produce seed and regeneration is established. At present, a polycyclic system is being employed on a cycle of 40 years and an assumed rotation of 80 years for most species. A very wide range of species is accepted in regeneration as desirable, and few silvicultural treatments (such as climber cutting) are now considered necessary. Essentially, the current recommendations are seeking to replicate natural gap creation and regeneration processes, and involve lower rates of removal and the establishment of smaller gaps than in earlier silvicultural systems.

Neotropics

There has been little attempt at developing silvicultural systems in the vast rain forests that cover large parts of the Amazon and Orinoco basins. Conversion logging and selective logging over unmanaged areas are widespread, but sustainable yield management is only an objective in small-scale research projects.

Suriname and Guiana are leaders in this respect. The Celos polycyclic silvicultural system, with a felling cycle of around 20 years (Jonkers, 1987), has been developed in Suriname. In this system, the forest is characterized by slow-growing, shade-tolerant heavy hardwoods. Commercial species above 50 cm in diameter are selected, but very few trees are felled at each cut, and felling and extraction are planned to minimize damage to the forest. Since there are few light-demanding timber trees to fill large gaps, if felling were increased, the forest would soon degenerate into a mixture of low-value pioneer tree species, for example *Ochroma* (balsa), *Cecropia*, and woody climbers.

Much effort has been directed to reducing the damage during logging operations, and it has been found that care taken over these operations does not increase costs. Hendrison (1990) recorded the following felling damage percentages (assessed in terms of gaps in the forest resulting from felled trees and by the number of damaged trees).

- 14% with conventional logging, removing 8–10 trees ha^{-1}, or 20 m^3 (figures as high as 30–70% have been recorded in many other parts of the tropics).
- 8% with controlled logging, removing 11 trees ha^{-1} and a similar (20 m^3) volume.

The tending operations, or 'refinements', consist of climber cutting and the removal of competing (or potentially competing) non-commercial species. This means poisoning with glyphosate all competing trees above 40 cm trunk diameter and, in well-stocked forest only, poisoning any non-commercial trees above 20 cm in diameter situated within 10 m of a valuable tree.

CONCLUDING REMARKS

In the tropics, systematic silviculture is a relatively recent introduction, and only in very few regions do practices go back more than one or two generations of trees. From a silvicultural viewpoint, however, Wyatt-Smith (1987) considered that 'the techniques to practice natural management are largely known or could be modified to cover any, if not most local circumstance'. There is probably enough scientific knowledge already available to make silviculture and conservation work successfully in the tropics. As Palmer and Synnott (1992) commented, 'It is one of the many paradoxes of tropical forestry over the past 30 years that the rise in public interest has been paralleled by a decline in the application of systematic management. It is also paradoxical that the same period has seen a great increase in research on tropical biology but little corresponding incorporation of research results into management practice'.

Foresters continually have to choose between different silvicultural systems to achieve different mixes of products and benefits from specific forest areas. No single system is ideal for all situations. The choice is most often between even-aged monocultures, which are usually based on planting with clear-cutting, and systems based on natural regeneration, which are, to various degrees, uneven-aged.

Stands of irregular structure and tolerant (shade-bearing) species are best suited to uneven-aged silviculture. This type of silviculture is also best practised on fragile sites, steep slopes, sites with high water tables, and very dry sites, which would be adversely affected by complete removal of the forest cover, even for short periods. Even-aged systems are most appropriate in stands of intolerant (light-demanding) species, and should be used to return over-mature, decadent, diseased, or insect-infested forest stands to productivity.

CHAPTER 12
Tree pruning and surgery in arboriculture

David Thorman

INTRODUCTION

Tree surgery is an important aspect of arboricultural care, and much of the work of the modern professional tree surgeon involves cutting with either handsaw or chainsaw (**521**). Ideally, such human interventions should not be necessary and trees in the wild have their own survival strategies; various species of broadleaved trees such as *Quercus* (oak, **522**) and *Olea* (olive) can survive for many hundreds of years, while the Californian coniferous redwoods (*Sequoia sempervirens* and *Sequoiadendron giganteum*) generally live for much longer (Chapters 2–3).

521 Tree surgeon at work removing a branch high up on a mature specimen of *Fagus sylvatica* (beech).

522 Ancient specimen of *Quercus petraea* (sessile oak, probably dating from the 13th century) still surviving at Cadzow, Scotland. (Photo copyright of Bryan Bowes.)

However, today more and more trees are integrated into the urban environment and suburban landscape. Consequently, they are found in city parks and arboreta, town centres, streets, and private gardens (**523**). Trees have been 'tamed', and the natural processes of growth and decay that take place, often unnoticed, in their wild habitats can become a problem in a built-up environment. When branches, boughs, or even the whole of a large and mature tree become unstable (**524**), their fall can have potentially disastrous consequences for humans in the locality (**525**).

All trees are subjected to the two principal forces of wind pressure and gravity, but the effect of these forces on a tree is dependent on its species, shape, size, leaf area, wood quality, location, and rate of growth. Throughout its life, the tree adapts itself to such forces and grows extra wood (reaction wood) in areas under most stress (Chapter 6). However, weaknesses and injuries do occur in the tree, either

523 Large *Quercus petraea* (sessile oak) amenity tree growing in suburban England.

due to natural processes or by man's activities. Tree surgery seeks to remedy these defects, usually by some form of pruning (*Table 13*).

Table 13 Tree defects and corrective pruning procedures

Defect	Possible remedial action
Rips caused by storms or branch removal (**526**)	Prune as close as possible to correct pruning position, and clean jagged edges
Cracks in branches	Remove or reduce branch
Cracks in trunk (**527**)	Reduce crown or fell tree
Dead wood (**528**)	Remove dead branches. Prune to correct position (Only if the dead branch is a safety issue)
Stubs	Prune to correct pruning position
Tight forks (**529, 530**)	Cable and rod bracing, or reduce crown
Decay in branches	Prune decayed parts or whole branch
Decay in trunk, bole, or roots (**531**)	Reduce crown (**532**) or fell tree
Loss of wood strength, subsiding branches, poor structure, branch architecture	Re-shape crown, reduce length of branch (**533**)
	Support branches
Dense crowns, rubbing and crossing branches (**534**)	Clean crown. Thin crown (**535**), but thinning may result in greater movement of remaining branches

524 Old specimen of *Quercus petraea* (sessile oak) hanging perilously from the rock face over a track around a country park in Scotland – the underlying path has been fenced off to keep walkers out of danger. (Photo copyright of Bryan Bowes.)

525 Loss of major limb in a mature specimen of *Aesculus hippocastanum* (horse chestnut) due to a weak branch union and decay.

526 Rips and branch stubs left after incorrect pruning of *Cupressus macrocarpa* (Monterey cypress).

527 *Acer pseudoplatanus* (sycamore) showing a longitudinal cambial 'rib' on its bark, indicative of an internal crack.

528 The long dead stub of the branch on this *Acer pseudoplatanus* (sycamore) is preventing the occlusion by callus of the wound at its base.

529 Large *Acer pseudoplatanus* (sycamore) trunk showing a tight fork with unequal stem diameters above.

530 *Fraxinus excelsior* (ash) trunk showing a very tight fork with a split.

531 Mature *Fraxinus excelsior* (ash) showing a large decay cavity below a tight fork in the trunk.

532 Crown reduction of a young horse chestnut. (Photo copyright courtesy of Roy Finch.)

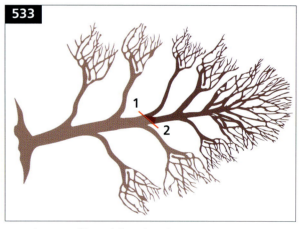

533 Diagram of branch length reduction on a tree. This branch is to be reduced by the removal of its tip and mid-region by a pruning cut, as shown in the line at position 1–2.

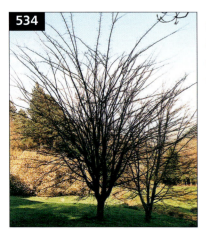

534 *Acer davidii* (snake bark maple) showing a very dense crown.

535 Diagrams showing a mature deciduous tree in winter before (left) and after (right) crown thinning.

RATIONALE AND PROBLEMS OF TREE SURGERY

The amount of pruning required to deal with a defect in a tree naturally depends on its severity; however, in addition to such remedial work, much pruning is undertaken to meet human requirements rather than to treat an actual tree defect. The main reasons for tree pruning are outlined below.

SAFETY

Falling trees, or their boughs and branches (525), can cause injury or damage, and a defective tree in a busy town poses a much greater risk than a similar tree on the verge of a remote country lane. Also, an ancient but defective tree may be in an isolated situation (522), or the public can be diverted around a similar important old tree occurring in a populated area. Hence, it is important for local governments to appreciate that some defective trees do not automatically pose a risk to the public.

HEALTH

Removal of diseased or damaged tree parts may prevent spread of disease, since the breakage of a bough or branch in a storm creates a large wound (525) which is open to infection. Also, the removal of damaged parts may prevent further damage to the tree.

OBSTRUCTION

Facilities such as power lines, roads, paths, and buildings may be blocked and this is probably the most common reason for tree pruning. Obstruction to power lines is of particular concern (536), since if branches come into contact with high-voltage conductors, the tree itself can become electrically energized. Low branches over roads and footpaths impede access and obstruct vision. Crown lifting (537) is used to improve access. Trees reduce incident light levels on adjacent buildings, while falling branches can cause structural damage to the superstructure.

SPACE

Trees which as saplings fit into the space of a small garden may at maturity become too large, as occurs with species of such trees as *Fagus* (beech), *Tilia* (lime, linden), and *Fraxinus* (ash).

TRAINING

Trees can be trained to achieve a desired shape or crown structure (532), while branches can be pruned to direct their growth away from an obstruction.

When pruning is carried out, it must minimize injury to the tree, and the various likely biological and mechanical effects of corrective surgery need to be considered.

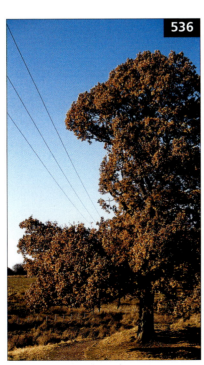

536 Large *Quercus petraea* (sessile oak) growing in Scotland. Several large branches have been cut away from this specimen so that the power lines above do not come into contact with the tree during stormy conditions. (Photo copyright of Bryan Bowes.)

537 Diagrams of a mature deciduous tree in winter before (left) and after (right) crown lifting.

- Pruning removes a proportion of the foliage in which photosynthesis would otherwise occur. Excess pruning can reduce the energy reserves of the tree, so that its health and vigour are adversely affected. The degree to which a tree reacts to this stress is also related to its health and pre-existing injuries, such as may have been caused by damage from adjacent construction work. With extreme over-pruning (538), a tree will decline and may eventually die.

- Branch or bough removal alters the shape and architecture of a tree, which previously had been adapted to its own local environment. Consequently, a sudden change in its wind loading can result in unaccustomed stress to the tree. One of the basic response mechanisms of a tree is to grow adaptive (reaction) wood in areas of weakness, such as at points of leverage. However, a tree which has undergone recent pruning (532, 536) may not have had the time or resources to cope with subsequent sudden environmental changes.

- Every pruning cut or tear on a tree leaves a wound (528), and the exposed vascular and cork cambia are often colonized by fungal or bacterial pathogens. A pruning wound can develop into a decay cavity, or even amalgamate with a vertical column of decay caused by several pruning wounds to the same branch.

TREE PRUNING TECHNIQUES

A knowledge of tree biology and tree mechanics should guide the tree surgeon when deciding which branches to prune. Trees have a natural defence compartmentalization system, which sets up barriers against the spread of decay (Shigo, 1984; Thomas, 2000; see also Chapter 6). The tree surgeon endeavours to work in association with this system when deciding on which parts of a tree require pruning and when choosing the correct pruning procedures (*Table 14*).

The position of the cut on a branch is a major factor in the prevention of subsequent decay at the wound, since protective phenolic substances are present in the wood at the swollen junction between stem and branch (branch collar, 542). Branch collars occur on all trees (539–541), although they are not all as clearly defined as shown for clarity in these diagrams. It is essential that the correct 'target' pruning position is selected (539); this should lie at the top and bottom of the outer margin of the collar (540, 541). Such a cut also

538 Savagely topped specimen of *Betula pendula* (silver birch) growing in a small garden in Scotland. (Photo copyright of Bryan Bowes.)

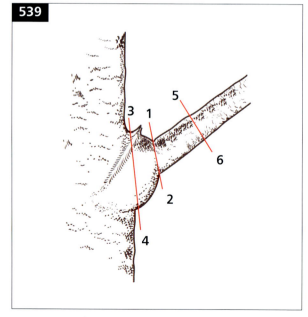

539 Diagram showing the correct position (1–2) and incorrect positions (3–4, 5–6) of final pruning cuts on a tree.

Table 14 Selection of the appropriate pruning procedure

Type of pruning	Reasons
Removal of one or more individual branches	A defect An obstruction by one or two branches
Crown reduction: can be an overall reduction in size or height, or of branch length on all sides or only one. May also refer to a reduction of foliage volume or branch weight (**533, 535, 537**)	To minimize sail area of crown in order to reduce leverage and stress on a weak point Insufficient space to grow Obstruction
Crown cleaning	Removal of dying, diseased, broken, or cracked branches
Crown thinning. Removal of branches within the crown and opening it out. Includes all those reasons detailed in crown cleaning, and also rubbing and crossover branches (**534**)	Reducing wind resistance Health and structure
Crown raising or lifting. Removal of lower branches up to a certain height (**533**)	Obstruction, access Light penetration
Shape and structural pruning which incorporates all of the above procedures	Pruning for the best architecture in relation to the setting

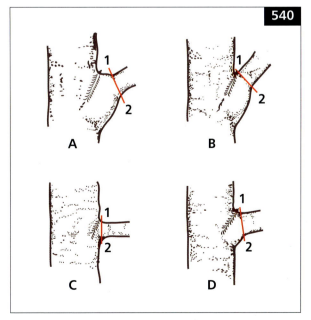

540A–D Diagrams showing that the correct pruning cuts (1–2) on a tree depend upon the position of the branch bark ridge and collar.

541 Sapling of *Salix* sp. (willow) showing a main stem, the branch of which has been removed by a correctly positioned target cut. (Photo copyright of Bryan Bowes.)

retains a protection zone at the base of the branch (542, right; 543), the wood of which contains phenols in broadleaves and terpenes in conifers, which resist the spread of decay. However, this zone is lost if the collar is removed by an incorrect cut made flush with the bough or trunk, and decay then spreads into previously uninfected wood (542, right).

When a pruning cut is made – or in wound where the bark is stripped – callus grows around the edge of the wound (see also Chapter 6). In a target-pruned branch cut, a regular ring of callus woundwood forms around it (543, 544) and may eventually completely cover the wound (545). In an incorrect flush cut, the callus forms mainly at the sides, with little or none at the top or bottom of the wound (546), and decay is often initiated at the base of the wound. To avoid any bark ripping at the wound, a branch should be cut into small sections during its removal (547). If stubs are left (526, 548), they will either produce sprout growth or die (528). If sprouts are produced, they may have weak attachments and become associated with decay in the stub (see pollarding below). A decayed stub will remain as a microbial-colonized site, and woundwood

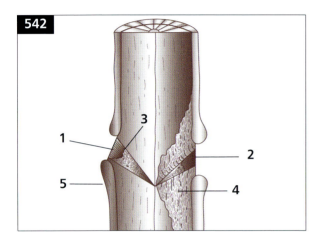

542 Diagrammatic radial longitudinal section along a young tree main stem showing wound callus (5) development after removing a branch. At the target cut (1) with an intact branch collar, a protection zone (3) is present, which extends from the surface some way down the branch base lying within the main stem. However, the incorrectly positioned flush cut (2) has removed the branch collar, and no protection zone has developed. Consequently, an extensive but irregular wedge of decayed wood (4) extends above and below the wound. Woundwood callus forms around the cut.

543 Callus woundwood has formed around a branch cut.

544 View of a pruning wound (*cf.* **543**) in the trunk of *Fraxinus excelsior* (ash) showing the well-developed protection zone.

545 Fully occluded branch scar on the main trunk of *Aesculus hippocastanum* (horse chestnut). (Photo copyright of Bryan Bowes.)

formation is obstructed (**528**).

There are advantages and disadvantages to all the procedures for tree surgery listed in *Table 14*. Too much thinning and opening up of the tree crown can induce more movement in the individual branches and cause them to crack or break, while 'top loading' (sometimes called 'lion tailing') is caused by removal of excessive inner canopy foliage, while retaining dense foliage on the branch ends. This effect can also be induced on the main trunk if most of the lower branches of the tree are removed.

Pruning reduces photosynthesis and can result in dysfunction (**538**). In general, therefore, no more than 30% of total canopy volume should be removed from a tree at any one time, and the same principle applies to individual boughs and branches. Photosynthates produced by the foliage on a branch may be stored in the trunk and roots, but a branch does not receive sugars synthesized elsewhere in the tree. Some tree species react strongly to over-pruning or 'topping' by growing profuse epicormic sprouts (**549**), or by developing stump sprouts or root suckers after felling to compensate for lost photosynthetic capacity (see also Chapter 6).

546 Trunk of a young *Prunus avium* (cherry) showing a wide pruning wound with most woundwood callus forming at the sides, while no callus has grown at the top of the cut.

547 Diagram of a young *Juglans* sp. (walnut) stem showing the correct way to prune off a well-developed branch. First, partial, undercut (1); second, partial, topcut (2); so the branch breaks between these two nearly adjacent cuts. Final target cut (3).

548 Badly pruned trunk of *Alnus* sp. (alder) showing a number of small side branches which have been incorrectly pruned to leave stumps instead of target cuts. (Photo copyright of Bryan Bowes.)

549 Specimen of *Tilia* (lime tree) showing dense epicormic branch growth, with many tight junctions, after topping.

PHENOLOGY

Many factors are involved when deciding the appropriate time to prune. The main consideration is usually the availability of carbohydrate reserves required by the tree to promote its defensive reactions after pruning. The summer, when photosynthesis is at its peak, would therefore seem to be a good time to prune, but pathogenic micro-organisms are then very active. In principle, the autumn, when food reserves have been built up by the tree, should be a suitable time. However, increased cavitation in the sapwood of the tree (Chapter 6) provides very favourable conditions for colonization by the profusion of newly released fungal spores. On balance, therefore it is better not to prune in the autumn.

The risk of microbial infection diminishes during the winter. Providing that freezing temperatures are avoided, this can be a suitable time to prune. Later, in early spring when bud burst (flushing) occurs in a temperate tree, there are high energy demands on it to allow new foliage growth; so this period is probably the least favourable time for pruning. The architecture of a deciduous tree crown can be seen clearly in winter (535, left) and allows the identification of those branches and branch unions which need pruning (535, right). The summer months generally provide suitable conditions for tree pruning, but the density of the foliage has a masking effect, which may make it difficult for a tree surgeon to assess where crown reduction or thinning is required.

In summary, therefore, midsummer and the end of winter seem the best times for pruning. Nevertheless, tree surgeons may be required to work at any time of the year, and healthy trees should be able to tolerate moderate amounts of pruning in all seasons. In injured trees, or those in ill-health, pruning in mid-spring could result in their decline and death.

POLLARDING

This technique is a widely practised feature of many street trees in Europe and elsewhere. The purpose of pollarding in such situations is to restrict the crown growth of potentially large tree specimens. However, it was previously a common technique in Europe (and is still used elsewhere; Chapter 11) to produce poles or larger timbers from trees growing in pastures, where the new growth would be out of reach of browsing animals (550). Pollarding involves a cut(s) of the central stem or trunk of a young tree at about 2–3 m above ground level. A new crown will form from the dormant buds sprouting, and the process can be repeated at intervals on the newly formed branches. When pollarded trees are re-cut, it should be above rather than at or below the pollard head (551). There may be decay at the original cut and, if the pollard head is intact, cutting into it can

550 Veteran pollarded specimen of *Quercus* sp. (oak), which is many hundreds of years old and is growing in an ancient wood pasture in England. (Photo copyright of Bryan Bowes.)

551 Old pollard head of a specimen of *Fraxinus excelsior* (ash) growing in England. This tree has been pruned many times – note the new branches arising from the recent cuts at the top of the head. (Photo copyright of Bryan Bowes.)

accelerate decay. Various species of *Salix* (willow), *Tilia* (lime), *Acer* (maple), *Fraxinus* (ash), *Ulmus* (elm), and *Platanus* (plane) sprout profusely, and in temperate regions are often the preferred species for pollarding.

Pollarding is often confused with the practice of removing, sometimes indiscriminately (**538**), the top of a mature tree (topping).The same procedure on major boughs is called lopping. In many species this results in abundant growth of epicormic sprouts (**553**), but the large wounds created may not become occluded and the procedure is often accompanied by decay. In topped trees, a pollard head does not easily form, but if a stable new crown forms it should be left to grow on. *Quercus* spp. (oak) trees are suited to such treatment but the new crown seldom attains its original proportions.

Species of *Tilia* (lime) and *Platanus* (plane) trees often successfully survive such treatment, but problems with decay and tight, unstable unions frequently develop and necessitate further cutting of the tree. For safety and conservation reasons, as in some veteran species of *Quercus* (oak) and *Carpinus* (hornbeam), it may be acceptable to top and lop a mature tree when a correct pruning procedure is not feasible. If such an old tree has developed a stable new crown, future maintenance pruning can follow the same guidelines as for a maiden tree.

ARTIFICIAL SUPPORT SYSTEMS

The tree has its own responses to weaknesses, but some additional mechanical support is often provided by the tree surgeon, either temporarily or permanently. Such measures include bracing, propping, and guying. Propping is not widely used but, as its name suggests, a heavy lateral branch is supported by a prop lodged between the branch and the ground (**554**). Guying is used for young trees that

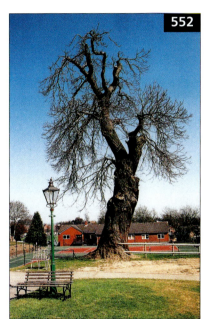

552 This mature specimen of *Aesculus hippocastanum* (horse chestnut), growing in a public area in England, has retained some shape following heavy reduction pruning because of extensive storm damage.

553 Stub with epicormic sprouts after removal of a large limb on this *Acer pseudoplatanus* (sycamore).

554 Ancient specimen of *Juglans regia* (common walnut), which still bears fruit, supported by a prop, and growing in the garden of Dumbarton Castle, Scotland. (Photo copyright of Bryan Bowes.)

have insufficient root anchorage. Bracing with cables or rods has been used extensively to prevent specific types of mechanical failure. Rods are inflexible and now rarely used, but cable bracing (**555**) provides support and allows some movement. This is important, because a rigid system restrains movement to such an extent that adaptive growth by the tree is not stimulated. Bracing with cables can help to prevent weak forks splitting apart. Stems with a tight fork (**529, 530**) and a weak union (possibly with included bark) are particularly suitable for this application.

For many years a cabling system using high-tensile galvanized steel wire has been used. A single cable consists of a length of wire rope of appropriate diameter, two eyebolts, and at least six wire rope grips. A hole of the diameter of the eyebolt is drilled through the branch, after which one eyebolt is inserted and fastened with a nut and round washer, and the bark is cut to match the placement of the washer. The other end of the wire rope is similarly secured, and the cable is then tightened with a ratchet, just sufficiently to take up any slack, and fastened at each end with three wire rope grips. In the case of multi-stemmed crowns, multiple cables in a 'box' system can be installed. Recently, a bracing system with polymer belt and rope attachments, which avoids the need to bore holes in the tree, has been widely used in continental Europe and the UK. This consists of polymer belts wrapped around the branches and attached, through two metal eyes, to a steel or polymer rope. An elastic insert can be part of the system as a shock absorber, ensuring that there is still some movement in the branches.

TREE FELLING

Much of the work of a tree surgeon lies in the removal of hazardous branches high up on a mature tree (**521**), and sometimes a dangerous large tree situated in a confined space has to be removed without causing damage to surrounding buildings. To dismantle such a tree is a highly skilled operation, involving a climber assisted by one or more ground persons. The tree is taken apart in separate pieces, with the climbing arborist making cuts with a small chainsaw, starting with the lower branches and eventually stripping the tree (**556**). The surgeon then works down from the top, gradually 'topping down'

the trunk or stems. Where space and circumstances permit, it may be possible to allow the cut timber to fall to the ground. Alternatively, parts of the trunk and branches are carefully lowered to the ground on ropes (**557**). A number of devices can used in branch lowering, including pulleys, steel karabiners, rope locking capstans, slings, and low-stretch ropes. When dismantling a large tree in more open access situations, it may be possible for the surgeon to work from a mobile platform raised from a well-anchored vehicle.

CONCLUDING REMARKS

Some mature broadleaved deciduous trees popularly used in arboriculture, such as species of *Quercus* (oak), *Fagus* (beech), *Fraxinus* (ash), *Aesculus* (horse chestnut), *Acer* (sycamore/maple), and *Castanea* (sweet chestnut), can be very long-lived and grow into large specimens. Hence, professional tree surgeons and arborists need to be sensitive to the needs of the tree; their main aim should be the safe care of these specimen trees, the lifespan of which is often several times greater than that of the average human. Tree surgery equipment is now highly sophisticated, and the professional arborist must train and become highly skilled in its use. Although legislation in some countries such as the UK can be rather restrictive, the tree surgeon must work within the various safety regulations, since pruning trees can often involve working at great heights (**521**) on sometimes weakened trees.

555 Specimen of *Aesculus hippocastanum* (horse chestnut) with its branches supported by cable bracing. (Photo copyright of Roy Finch.)

556 Tree surgeon at a preliminary stage of dismantling, removing the branches of a specimen of *Chamaecyparis lawsoniana* (Lawson cypress), which had grown too large for its garden in Scotland. (Photo copyright of Bryan Bowes.)

557 Dismantling a large specimen of *Fraxinus excelsior* (ash) – lowering a large segment of its trunk to the ground.

CHAPTER 13

Tree propagation for forestry and arboriculture

Brent H McCown and Thomas Beuchel

INTRODUCTION

The professional tree propagator needs to be a keen observer of plants, understand biology, and be a skilled technical practitioner. Propagation is so basic to plant science and agriculture that there is an International Plant Propagator's Society devoted to this subject. For the forester/arboriculturist/horticulturist, propagation is the path for perpetuation of a particular selection or species of tree or other plant. Successful propagators work in concert with the inherent capacities of a plant to multiply itself, capacities which have been honed through centuries of evolutionary selection. It is often much easier to complement the modes of propagation observed in a plant's ecological niche than to force into production a mode of propagation foreign to that species. Nevertheless, man's ingenuity has given us an array of technologies whereby we can extend the modes of propagation well beyond those expressed in the natural environment.

Today, a crop will not be a commercial success without an efficient means of propagation accompanying its production. There is a plethora of books and articles (Macdonald, 1986; Brown, 1999; Hartmann *et al.*, 2002), and now the internet, which describe propagation practices in detail. Repetition of that material would be both beyond the scope of this chapter and redundant. Instead, this chapter will attempt to give a background perspective in enough detail to allow the reader to make appropriate decisions as to the logical propagation practices to employ. In addition, emphasis will be on common problems encountered in propagation, and how to either avoid or overcome them.

OVERVIEW OF CONCEPTS

Trees can be propagated using either sexual or vegetative (asexual) modes. Sexual plant propagation always involves the formation of an embryo, following the fertilization of the egg in its ovule by a male gamete, usually culminating in the production of a seed. Depending on how tightly the genetics is controlled, the progeny (seedlings) may or may not greatly resemble the parents. Except in very rare instances, asexual propagation does not involve gametes and thus the progeny are generally an exact genetic duplicate (clone) of the original plant. In a few cases (apomixis and somatic embryogenesis), the two modes are combined, and an asexual embryo is produced in a seed or seed-like structure.

In tree propagation by either sexual or vegetative means, controlling the genetics is of paramount importance, because it is the genetic make-up of the progeny that will determine the diversity and the value of the propagated population. With clones, the genetic diversity of the progeny should be zero – all the individuals will have an identical genotype to the parent (ignoring possible chance genetic alterations like mutations). In such cases, the phenotype of the vegetatively propagated progeny (growth rate, disease resistance, flower/fruit characteristics, etc.) will reflect that of the parental genotype, subject to environmental effects.

For sexually produced seedlings, the progeny can be essentially identical to the parents if highly controlled breeding systems, like pureline and inbred-generated hybrid cultivar production, are employed. Purelines and inbred-generated hybrids depend on obtaining genetic homozygosity through generations of inbreeding. However, for trees, such intense inbreeding is not generally tolerated biologically, and thus seedling populations of trees usually vary widely in the diversity of their genetic make-up. Most often a propagator desires to have some control over this diversity, so that the general characteristics (adaptability, commercial value) of the seedling population are predictable. Limiting seed collection from trees to defined regions (provenance, seed collection areas, seed orchards) can put bounds on the degree of genetic diversity in a seed source.

Quite often, the inherent propagation capabilities of a tree can be predicted by studying its ecology. For example, species of such trees as *Eucalyptus*, *Populus*, and *Salix*, which are pioneers of land not currently dominated by trees, will often produce very large quantities of small, highly mobile seeds in the spring. These species have relatively short life cycles and begin producing seed within the first decade of their lives. Such seeds may have little food reserves or protective mechanisms, and thus tend to be short-lived, but will germinate quickly after dispersal. In addition, these tree species will often have aggressive modes of vegetative propagation, such as root suckering or rapid rooting of buried branches. In contrast, trees characteristic of more mature (climax) forest niches, such as species of *Swietenia* (American mahogany) and *Tectona* (teak), often produce smaller quantities of much larger seeds. These tree species have longer life cycles and display no prominent asexual propagation modes. For the plant propagator, the pioneer tree species are generally much more readily propagated than the latter examples.

A second useful aspect of the ecology of a tree species is to study the tree's regional environment. For example, trees which evolved in regions that undergo yearly seasonal stresses (cold, drought) will have mechanisms that time their reproductive biology to avoid these stresses. Thus, trees native to regions experiencing months of freezing winter temperatures generally have seed dormancy mechanisms to delay germination until spring. Understanding these mechanisms is critical if controlled, large-scale propagation of such trees is desired.

One aspect of tree biology, the plant's life cycle, dominates propagation practices, whether sexual or vegetative. The life cycle can be roughly divided into the juvenile and adult phases. Tree species differ widely in the time it takes to progress through the juvenile into the adult phase, with some taking years and others many decades. During the juvenile phase, trees will not produce seeds even if given appropriate environmental cues. However, vegetative reproduction techniques are generally much more successful using juvenile tissues. As a tree enters the adult phase, seed production can occur, but the capacity for vegetative propagation dramatically decreases. Interestingly, with many species, a reversal of phase from adult to juvenile (rejuvenation) can be stimulated, and thus the capacity to be vegetatively propagated regained. The control of juvenility is one of the major tools of the tree propagator.

PROPAGATION BY SEED

If a system for propagating plants were to be designed, it would be difficult to perfect something more ideally attuned to fulfil this purpose than the seed (Iriondo and Perez, 1999). Protected by various coverings, the embryo is wrapped in nutritious tissues and can tolerate all sorts of environmental extremes. Given the right conditions, some seeds can remain viable for thousands of years, such as those from large-seeded, hard-shelled leguminous plants. In other circumstances, the process of germination commences, leading to absorption of water (imbibition), swelling and emergence of the radicle (embryonic root), solubilization of stored food, resumption of cell division, and finally the appearance of the leafy shoot. A propagator complements this dynamic process by providing both stimulatory treatments and the appropriate environmental conditions.

The first step in propagation by seed is the procurement of the best quality seed. Quality can be measured in both genetic and physiological terms. Genetic quality will be in large part defined by the ultimate destination of the propagated trees. If the project is ecosystem restoration of degraded or disturbed lands, then seed quality will include components of genetic diversity – the seed needs to be as representative as possible of the natural community being restored. However, if the trees are part of an urban forestry project and destined for street plantings, then resistance to common urban stresses such as compacted soils will define the required genetic quality. In any case, adequate time spent in obtaining the best seed for the intended purposes will be very well rewarded.

The best genetic make-up will be of little value if the physiological quality of the seed is not also preserved. Embryo viability and amount of stored food reserves are the main components of physiological quality. While such quality is already established during the seed production, proper storage conditions will maintain seed quality. Although specialized seed storage conditions are required for some tree species (such as some nut species), in general a cold, dry, and stable environment is ideal for most seeds being stored for a few years. Seed storage can be critical for many tree species, as often seed-bearing trees produce large quantities of seed during only one or two years out of every five-year period.

The factors important in the germination environment include water availability, temperature, adequate aeration, and freedom from pathogens. Well-aerated germination media and a constant moisture supply benefit almost all seeds. The appropriate germination temperature varies with the tree species. In addition, some species may benefit from special conditions, such as exposure to light or alternating temperatures.

One aspect where the propagator can exert considerable influence on germination is through pretreatment of the seed. Many tree species have intricate mechanisms to ensure that germination occurs in synchrony with the growing season in which they evolved. Various pretreatments can mimic environmental cues and thus stimulate more uniform and consistent germination. By knowing the native climatic region and specific seasonal environment, it is often possible to anticipate what pretreatment is likely to be required. Several pretreatments are commonly practised.

- Seed cleaning. Coverings on the outside of the seed coat may contain substances, usually phenolics or plant hormones, which can inhibit germination even if the seed is placed under ideal conditions. Often seeds covered by fleshy fruits (for example, some palms, and members of the Rosaceae) will not germinate uniformly in nature until the fruit wall is fully removed by either passing through a herbivore or decaying in the soil. Thus, in seeds used for propagation, adequate cleaning of the seed is important. Such fleshy coverings can be conveniently removed by submerging the fruits in water (558) and allowing micro-organisms to remove the pulp through fermentation. The viable seeds will sink to the bottom of the container, while the pulp and other chaff will be carried to the top in the foam. With dry fruits, forcing through screens is a common cleaning practice.

- Warm stratification or after-ripening. Although relatively rare with trees, with some species such as *Fraxinus*, the embryo is not fully mature when the seed is harvested or shed. A warm (20–25°C) and moist treatment will allow the embryo to increase in size and mature.

558 Newly harvested fruits of a member of the Rosaceae floating in a large vat of water. The mixture will be allowed to ferment, and the fleshy fruit covering will be degraded. The freed and viable seeds fall to the bottom, while the pulp, debris, and empty seed float among the foam. Consequently, the process both cleans the seed and selects for viability.

- Scarification. The first step in the germination process is the absorption of water (imbibition). This is prevented in some seeds by the presence of an impermeable layer, usually associated with the seed coat. As long as the layer remains completely intact around the seed, germination will not occur. Scarification physically disrupts this layer and provides pores where water can enter; the subsequent swelling of the seed completes the process of disrupting the impermeable layer. In nature, scarification occurs through digestion in an animal's gut, freezing/thawing, micro-organism attack, or fire. Artificial scarification can be accomplished either by mechanical manipulation (such as exposing the seed surface to sandpaper), or by treatments with concentrated sulphuric acid for up to six hours. The latter treatments will also remove any substances in the seed coat that are inhibitory to germination.

- Hot water treatment. Some species regenerate best after a fire has swept through the native community (Brown, 1999). Seeds in the soil are stimulated to germinate by fire as a direct effect of the heat as well as by other possible changes. Many of these species have seeds that contain a layer of live cells just below the seed coat, which actively inhibits oxygen uptake. Seed germination involves active metabolism, and thus the oxygen demand is high. The heat of a forest fire will kill this layer of cells, allowing oxygen exchange. Hot water treatment can mimic this effect, but boiling water can kill thin-coated seeds and cold water for a longer period (for example, with tropical leguminous tree seed) may be just as effective and safer. Water treatment may also be effective in removing substances in the seed coat that are inhibitory to germination. Such a condition is commonly found with species native to arid environments, where seasonal rains would accomplish the same function.

- Cold stratification. Tree species native to temperate climates must synchronize their germination cycles with the winter stress period. The most common mechanism is the requirement for a cold period to overcome the dormancy of the embryo before germination can commence. In nature, this treatment is provided by winter, and thus germination is timed to occur in the spring, a period

favourable for seedling survival and growth. Seedling nurseries often utilize natural stratification in the field for large quantities of tree seeds (**559**, **560**). Under more controlled conditions, cold stratification can be given by exposing the seed to moist cold (around 5°C) conditions. Note that moisture and imbibition are needed, thus refrigerated dry storage of seed will not generally overcome embryo dormancy. This is because during the stratification, active metabolism gradually changes the internal hormonal balance in the embryo from one of inhibiting growth to one of promoting growth, and the embryo must be fully hydrated for this metabolism to occur.

559, 560 Planting of *Fraxinus* seed in the field in the fall (autumn). The seed is hand-sown on prepared beds (**559**), then rolled to make good contact with the soil and covered with clean sand (**560**). The seed will be stratified through the fall and winter months. This breaks embryo dormancy and allows germination in the spring.

- Complex treatments. It is not uncommon for a tree species to require a sequence of treatments to simulate the conditions promoting germination in its native environment. For example, scarification followed by stratification allows imbibition to occur, so that subsequent metabolism under cold conditions will overcome embryo dormancy. Likewise, after warm stratification, an immature embryo once matured may develop deep dormancy and thus require a subsequent cold stratification treatment before germination can occur.

VEGETATIVE PROPAGATION

A high genetic diversity of a tree population is a critical requirement for such purposes as restoration of a native community from seedlings. However, in many other situations, such as urban landscapes, a higher degree of control over the genetics (and thus a greater predictability of the characteristics of the plant population) may be required. For trees, the easiest methods of achieving a high genetic uniformity are through vegetative propagation or cloning of superior individual genotypes. A clone consists of the identical individuals of a single selection (genotype) that are maintained by vegetative (asexual) means. For trees, there are four applicable methods of cloning: cuttings, layerage, grafting/budding, and micropropagation.

CUTTINGS

The basis of a cutting is the regeneration of missing parts on a piece of the plant to be cloned. For trees, stem (shoot) cuttings are by far the most common cutting type. With a shoot cutting with buds, only the roots need to be regenerated and thus the propagator's focus is the formation of new root initials, termed adventitious roots, at the base of the cutting (**561**). With some tree species such as *Populus* and *Gymnocaldus*, sections of roots can be used as cuttings, and the regeneration must then involve the development of both adventitious shoots and roots. There is still a lack of understanding of the biological controls involved in organ regeneration in plants, and the techniques that a propagator uses have been derived from both scientific explorations and trial-and-error knowledge.

The factors important in adventitious rooting of cuttings are numerous and prominent among them are the following.

- Juvenility. Cuttings taken from stock trees that are in a juvenile phase of growth have a much greater capacity to generate adventitious roots than cuttings taken from adult stock. Thus, the maintenance of the stock plants becomes critical to successful stem-cutting propagation of trees. There are several common approaches for both 'rejuvenating' an adult tree (such as sequential grafting of adult shoots onto highly juvenile seedlings), as well as maintaining a clone in the juvenile state. Many of the techniques involve the collar region of the tree. The tree collar lies at the juncture of the root and shoot, and is a region which appears to maintain a high degree of juvenility throughout the life of a specimen. Hence, confining to near the tree collar the seasonal growth of shoots helps to ensure juvenile stock material for cuttings. One approach for maintaining juvenility is to seasonally shear the stock plants back to the collar region (hedging), thus forcing all new shoot growth to develop from there, either adventitiously or from pre-existing buds.

561 A well-rooted cutting of a deciduous *Euonymus*. The cutting was taken in mid- to late summer (semi-softwood) and stuck in rooting beds in a hoop house (background) under mist. After rooting, the cutting is fully dormant and must receive a cold treatment before growth will resume in spring.

- Timing. Different species and selections of trees can differ widely with respect to the most successful time for rooting cuttings during the seasonal growth cycle. The material used can vary from succulent seasonal flushes (softwood cuttings), to less succulent and more rigid new growth (semi-softwood, semi-hardwood), to hardwood cuttings not in active growth and which are leafless if deciduous (562, 563). The trend in propagation of tree cuttings seems to be towards the semi-softwood type, since this is still highly responsive to treatment but not overly difficult to maintain in a viable state during the rooting process (564).

- Wounding. The formation of adventitious roots is often associated with wounds on the stem of the cutting. In taking the cutting, a wound is produced at its base; however, additional wounding is often conducted, such as by removing a strip of bark along the side of a cutting, to promote additional adventitious roots.

- Hormone treatment. The plant hormone (auxin) is involved in the adventitious rooting process. By applying auxin treatments to the base of a cutting, the propagator can change its internal hormone balance, and thus stimulate adventitious rooting where it might not otherwise occur, uniformly or with suitable intensity. The most common auxin hormone utilized for tree cuttings

562 A bundle of hardwood cuttings removed from cool storage. The cuttings were taken in the early winter from the previous season's growth ('whips'), cut to a uniform length, and the tops sealed with tape to prevent desiccation. The basal ends (here covered with wood shavings) were wounded, bundled, and treated in a quick dip of rooting indolebutyric acid (IBA) hormone. The bundles are stored in a cooler, where callusing of the wounded areas occurs, and then planted in propagation beds in a cold frame, hoop house, or directly in the field.

is indolebutyric acid (IBA), and commercially available preparations, either as a talc base or as a liquid, are readily available.

563 Planting of hardwood cuttings directly in the field in early spring. Only relatively easy-to-root trees (such as *Populus*) are commonly rooted in this manner, as the ability to control the rooting environment is minimal under field conditions. However, this method is inexpensive and allows the newly rooted plants to grow in place for a full season.

564 A ground bed of semi-softwood cuttings of various shrubs and trees. This bed is in a hoop house, where high humidity levels can be maintained during rooting. The cuttings will grow in place until the fall or early winter, when they are harvested and stored or planted.

- Rooting environment. Keeping a cutting viable while adventitious rooting occurs is critical and, for all but some leafless hardwood cuttings, this involves high humidity. In the rooting environment this is usually maintained by periodic misting (**565**) or tenting (**566, 567**). In addition to humidity, temperature control is important, especially the maintenance of a uniformly warm temperature at the base of cuttings where rooting will occur.

LAYERING

This technique is analogous to that used for cuttings, except that the 'cutting' or 'layer' is kept partially attached to the parent plant during the formation of adventitious roots, and thus environmental control is not so critical or complicated. Layering is commonly practised with tropical evergreen tree species such as *Ficus* (fig) by using air-layering (**568**). Additionally, trees for which cuttings are very slow to root can be propagated by mound or trench layering, as the parent plant keeps the 'cutting' viable for this extended period.

GRAFTING/BUDDING

For trees destined for urban landscapes or orchards, the most utilized cloning technique has been graftage. In contrast to cuttings, grafting (the scion being a stem piece with multiple nodes) and budding (the scion being only a single bud) regenerates a whole plant by combining the parts of two or more different genotypes from the same genus. Commonly, the root systems (rootstocks) are seedlings. With fruit trees, clonally produced rootstocks are utilized, as these stocks have been selected to add important traits, such as dwarfing and disease resistance, to the grafted tree.

Several different grafting/budding techniques are utilized to propagate trees. For non-conifers, whip-and-tongue (**569**), and shield or chip budding (**570, 571**) are common. Side or veneer grafting is

565 Cuttings of *Pinus* stuck in plugs on benches in a screen house in Australia. The cuttings are taken from partially hardened new growth from sheared and hedged stock plants, in forest tree stock plant blocks in other areas, and transported under refrigeration to this facility. Note the mist being applied by a beam that moves up and down the rooting benches according to preset timing.

566, 567 Rooting of semi-hardwood conifer cuttings in a greenhouse in trays filled with a medium rich with the perlite soil amendment (**566**). After watering, the benches are covered with a clear polyethylene tent (**567**) to maintain humidity during the months required for the rooting process.

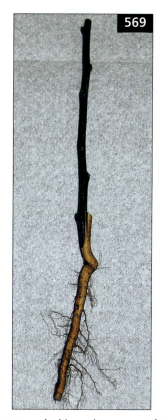

568 Air-layering of a variegated selection of a tropical tree (*Schefflera*) propagated for ornamental use. After stripping off the leaves along a section of stem, the stem is wounded and covered with a moist medium, often sphagnum moss, then wrapped in foil or plastic. After rooting, the stem is cut below the rooting, and the layered propagule is planted.

569 A newly prepared whip-and-tongue graft of a hardwood tree. The darker upper portion (scion) is the clone, and the lower lighter-coloured portion is the rootstock. The graft union will be wrapped (see 568) to prevent desiccation, and to tie the two members tightly together at their union.

570, 571 A field of newly budded rootstocks of *Fraxinus* (570). Buds of the scion selection have been inserted near the collar of the sapling rootstock – the white plastic covering indicates the budded area. Removal of this covering reveals the chip bud of the scion (571). After the graft union takes, the top of the rootstock will be removed just above the budded scion; this forces the scion to grow out and eventually become the new aerial tree.

572 A side or veneer graft of an evergreen (*Juniperus*). In this graft, the top of the rootstock (here a potted plant) is maintained until the graft union is complete. Such grafts are commonly used when the graft union process requires extended time. These grafted plants will be placed in a cool and humid location until the graft heals. The top of the rootstock will then be removed.

frequently employed with conifers (**572**). The timing of graftage depends on the environment. Grafting is most commonly done during periods of inactive (dormant) growth (**573**, **574**), whereas budding is usually conducted during the growing season.

If the tree to be cloned is not juvenile, then grafting is probably the best approach, as grafting is not highly influenced by the phase state of the tissues being manipulated. Under normal commercial conditions, the scion material is usually current or one-year-old growth, and the rootstock is a one-to-two-year-old seedling. This form of cloning is successful with most tree species, but some trees, such as certain species of *Quercus* (oak) and *Juglans* (walnut), are notoriously difficult to graft.

Besides poor technique, the principal problem encountered in grafting/budding is incompatibility between the grafted members, a phenomenon for which the biology is not well understood.

573, 574 A bundle of new whip-and-tongue grafts of a hardwood tree (**573**). The grafts are placed in crates with moist peat or coarse sawdust (**574**) and stored in a cooler. The graft union slowly heals, dormancy is broken, and the crate is removed from the storage area for planting out of the grafted plants.

Unfortunately, in some cases incompatibility reactions may not be evident until years after the graft was performed. This results in the premature death of trees of considerable size, thus amplifying the economic impact of the loss. For this and other reasons, tree clones – which were previously propagated by graftage – are now more commonly cloned using cuttings.

MICROPROPAGATION

The newest technique for the cloning of trees is multiplication of plants using sterile (*in vitro*) techniques, generally called micropropagation, where the term 'micro' refers to the relatively diminutive size of the tissues being handled. The advantages of micropropagation over other cloning techniques are that it can be done all year round, and that large numbers of micropropagules can be produced in a relatively short period of time from a limited supply of the stock selected to be cloned. Thus, micropropagation is often preferred when introducing a new selection to commercialization. With species where normal cuttings are very difficult to root (for example, species of *Amelanchier* (service berries), *Syringa* (lilac), and some *Eucalyptus* species) or inappropriate, as in palms and bamboos, micropropagation may be the only commercially successful method of cloning. Two different approaches can be taken to micropropagate a tree selection: shoot culture and somatic embryogenesis (Merkle, 1999).

In vitro shoot culture of a tree involves a number of stages of acclimatizing the shoot explant to the culture conditions and subsequently optimizing its growth. To microculture a tree shoot, the first step is the isolation phase (575), which involves freeing the explant (usually a young, rapidly growing shoot) of micro-organism contaminants, and stimulating its continued growth under microculture. The next step, stabilization, is the most unpredictable and involves continued and rapid subculture of the new shoot growth. As the tissue is subcultured, a change in quality of the new growth occurs from erratic, distorted, large-leafed shoots to uniform, small-leafed, continuously growing shoots (576). The time required to stabilize a plant varies markedly and may entail a few months or several years' maintenance in culture. The biological changes during stabilization are unclear; however, rejuvenation is obviously a major part of the change. After full stabilization, the microshoots often closely resemble young seedlings of the tree species concerned. The initial material for microculture can have a strong influence on success, with highly juvenile stock being much more readily established than adult material.

575 Common steps involved in placing a tree (*Populus*) into microculture (isolation stage). Shoots are divided into regions (buds, shoot tip, and nodes), sterilized, and placed into sterile vessels on suitable medium to initiate new growth. The proliferating shoots are repeatedly subcultured until their form stabilizes, and they are more uniform and seedling-like.

576 Stabilized and optimized shoot cultures of three trees (right to left, *Ulmus*, *Amelanchier*, and *Betula*). From such shoot cultures, succulent individual shoots (microcuttings) are harvested and placed in a propagation area for rooting and acclimatization.

Once stabilized, the next step in shoot culture is to optimize the growth and production of new shoots. In shoot culture, the mechanism of multiplication is in the stimulation of axillary branching through the use of the plant hormone cytokinin. A successful shoot culture forms bundles of rapidly growing shoots (576) through endless branching. During optimization, the specific hormones, their concentration, the mineral medium,

577, 578 Somatic embryogenesis of a conifer tree. Somatic embryos are induced to form on the surface of a specialized mass of cells (embryogenic callus, **577**). One approach to moving such embryos into production is to encase them in a nutritive 'artificial seed' (**578**). (Photos copyright of R. Durzan and D. Ellis.)

and the culture environment are all monitored with the goal of producing the greatest number of healthy, uniform shoots.

The next stage of shoot culture micropropagation involves removing the material from microculture and establishing complete young plants. The microshoots are harvested as microcuttings and are handled like very small softwood cuttings. Since the microcuttings are both softwood and juvenile, they generally readily regenerate adventitious roots. The microcuttings are very prone to desiccation, and so control of a uniform humidity in the rooting environment is critical. As rooting occurs, humidity is reduced and light intensity is increased to gradually acclimatize the plants to normal greenhouse conditions. The end result is a young, rapidly growing micropropagule which looks and acts much like a seedling of that species of tree.

The second approach to micropropagating a tree is somatic embryogenesis. Although this technique has not proven nearly as commercially useful as shoot culture, its potential is enormous. In somatic embryogenesis, the multiplication is based on the generation of adventitious embryos (**577, 578**). These somatic embryos closely resemble morphologically their seed cousins (zygotic embryos) except that they are clonal and not derived from the union of gametes. When successful, literally millions of embryos can be generated, with a minimum of labour, using liquid culture in fermentation-like vessels. However, two major obstacles have limited the application of this approach. Firstly, generating adventitious embryos from non-seedling tissues (i.e. proven clonal material) has been difficult with most tree species. Secondly, the conversion of somatic embryos into viable propagules capable of being planted out has been highly inefficient. Tree somatic embryogenesis has found its widest application with forestry conifer species, namely *Pseudotsuga* and *Picea*.

Micropropagation requires a considerable amount of skilled labour and unique facilities, and is therefore a relatively costly method for cloning trees. One approach to reducing the costs (as well as capturing some of the advantages of micropropagation) is combining micropropagation with other cloning methods. For example, micropropagation can be used to generate disease-free, juvenile, and rapidly growing stock plants, which are subsequently extremely useful for cutting propagation.

HANDLING OF YOUNG PROPAGULES

One of the lessons of the past few decades is that, no matter how well the initial propagation has been accomplished, mishandling of the young propagule in its final planting can result in inferior

579 Beds of pine seedlings in a propagation field. The seed was directly sown in these beds and the plants have grown for one season. Roots of such field-grown propagules will generally not be subject to root deformations, as may occur in container-grown propagules (see 583).

580 Young seedlings of *Picea pungens glauca* (blue Colorado spruce) germinated and growing in plug trays in a hoop house ground bed. The handling and design of the plug containers are important for minimizing root deformation problems.

581 Root deformation in a one-year-old *Ulmus* tree. This tree was dug from a nursery plot, but was graded to be destroyed because of roots encircling the base of the young tree. If this specimen was out-planted to a permanent site, as the tree grew over the years and the stem and roots increased in diameter, the stem would be increasingly girdled by the encircling roots, eventually killing the tree. The encircling roots were a result of improper handling during propagation.

performance of the tree. When seeds or cuttings are propagated directly in the field (579; see also 583), root deformations are rare. However, major problems can occur when the propagation is done in containers (565 and 580), a production system becoming the standard in many regions. Of particular concern is deformation of the young root system, caused by poorly designed containers in which the new tree propagule was grown. Tree roots are generally relatively coarse, and the main roots on a new propagule will grow larger in diameter for

the life of the tree. Any deformation of these roots will persist in the future tree and potentially interfere with its growth. The most common deformation is in roots which circle as they encounter the side of a container, and then continue to grow down the side of the container in a circling pattern and endlessly circle at its bottom. Later in the life of the tree, such circling roots may constrict and girdle the lower trunk (581), interfering in the vascular flow and leading to growth disruption and eventual death of the tree.

582 A container-grown conifer tree (on left) showing the quality of the root system. Note the vertical ridges that help to direct the roots downward when they contact the sides of the container. In addition, the container (on right) does not have a solid bottom and has been internally coated with a copper compound. Both of these features inhibit root circling and facilitate fibrous branching.

583 Seedlings of *Picea* grown under three different regimes: seeded in the field (left), seeded in a standard container (middle), and seeded in a specially designed plug tray (right). The plug tray was designed to achieve 'air-pruning' of the roots as they grew down the plug, thus promoting branching and a fibrous root system, which inhibit root circling while increasing transplanting success. Note also how the standard container altered the overall architecture of the seedling root system (including the beginning of circling of the roots that were at the bottom of the container) as compared to the field-grown specimen.

Encircling roots and other root deformations can largely be prevented by choosing a well-engineered propagule container. Such containers will have features to prevent root deformations.
- Corners or ridges along the sides (**582**). These entrap roots beginning to circle and direct the growth straight downward.
- No solid bottom to the container (**582**). If placed on a non-solid surface or raised above a bench surface, roots emerge from the bottom of the containers and encounter air, and the tips are killed. This effectively 'air prunes' the roots, prevents their encircling the bottom of the container, and also promotes more fibrous branching of the root system (**583**).
- Some newer containers have open slits running vertically down the sides, or even ledges with openings running around the container. These features also act to prune roots.
- In addition to the above design characteristics, containers may be treated with a copper hydroxide or copper carbonate internal coating, which will kill root tips as they contact the sides and bottom of the container.

No matter what species is being grown or how the initial propagation was conducted, it is the essential responsibility of the propagator to take measures to avoid deformed tree roots.

ETHICS IN TREE PROPAGATION

One of the unique challenges for the tree propagator is that trees live for many decades and have a dominant influence over the environment in which they are planted. The inappropriate choice of new trees can have extensive and lasting negative impacts on the areas planted, so it behoves tree propagators to seriously consider the most appropriate methods of propagation and the characteristics of the material being used. It is unethical for native restoration purposes, to knowingly produce tree seedlings from a seed source of limited genetic diversity or not adapted to that environment. In addition, to endlessly propagate a few clones for either urban forestry planting or massive regional plantings is also unethical. A very narrow genetic base in such clones not only increases the risk of pest/disease attack but also raises the possibility of inbreeding depression in subsequent generations of trees. The potential 'genetic vulnerability' of an artificial population of trees is under some considerable control, and the lessons learned from past mistakes (for example with Dutch elm disease) should be heeded by propagators.

Another ethics question concerns the proprietary rights of breeders and developers of new, improved genotypes of trees. It is becoming more common to have new tree selections covered by some form of legal protection, patenting and trademarking being the most prevalent. The propagator is usually the first line of information for growers of such restrictions. In addition, it is usually at the propagator's level that the legal protection is monitored. This means that the propagator should keep track of quantities of material propagated, its subsequent distribution, and collection of the royalties. Thus, it falls upon those conducting tree propagation to respect the rights associated with the legal protection and to ensure that the system functions properly. Ideally, such proprietary systems will lead to better promotion and development of more superior trees, a scenario where both the propagator and all others involved will benefit.

CHAPTER 14

Forest and woodland conservation

Ghillean T Prance

INTRODUCTION

The most important way to preserve trees is in natural forests and woodlands (**584, 585**). On site, or *in situ*, conservation is preferable to preservation in arboreta or in seed banks because it sustains the trees, and the many other organisms that depend upon them, in their natural environment. The interactions with pollinators and agents of seed dispersal are maintained in forests, which are also vital in many places to sustain the natural environment. Forests help to control climate and rainfall, and are often essential to conserve watersheds. In any natural ecosystem the process of evolution and adaptation to change takes place naturally.

At present approximately 8.8% of the land surface of the earth is within so-called protected areas, and of

584 Rain forest of the Rio Xingu region of Brazil (see text and *Table 15*).

585 Wood of deciduous *Fagus sylvatica* (beech trees) located near Callander in Scotland. (Photo copyright of Bryan Bowes.)

586 Amazon rain forest burning. In 2002, the second highest level of deforestation occurred in this region, showing the urgent need to increase conservation efforts.

587 A one-hectare experimental reserve in the 'Dynamics of Forest Fragments' project near to Manaus, Brazil.

588 Aerial view of one- and ten-hectare experimental forest reserves in the 'Dynamics of Forest Fragments' project near to Manaus, Brazil.

these roughly 30,350 reserves, only some are for forest ecosystems (World Resources Institute, 2001). A recent survey showed that only a small percentage of these could be considered secure (Dudley and Stolton, 1999). Also, the destruction of forests and other ecosystems that are not yet in protected areas (**586**) is accompanied by the extinction of some species. As outlined previously in Chapters 2–5, there are many different types of forest in the world, and it is most important to conserve as large a sample of these as possible. Some of the factors that need to be taken into consideration to conserve forests are discussed below.

RESERVE SIZE

Reserve size has been a much debated topic (Soulé, 1987). In particular, it has often been about the relative merits of a single large reserve versus several smaller ones. The rate of species loss and genetic deterioration for most species is inversely proportional to reserve size (Soulé, 1986, 1987). Small reserves are also more prone to complete destruction than large ones. A small reserve can easily be destroyed by a fire or a storm, whereas in a larger area the destruction is unlikely to be total.

An interesting experiment in reserve size is the 'Dynamics of Forest Fragments' project in rain forest north of Manaus, Brazil. Here, forest fragments of different sizes have been allowed to remain within an agricultural area (**587, 588**). Data collected over 25 years about many different organisms have been gathered in this project. An effect that soon became

589 Edge of a small reserve of the 'Dynamics of Forest Fragments' project near to Manaus, Brazil, showing a serious edge effect after only a few years of establishment.

590 The maintenance of pollinator relationships is an essential part of conservation. Here, a carpenter bee is entering the flower of *Bertholletia excelsa* (Brazil nut tree). (Photo copyright of A. Henderson.)

apparent in small fragments was the so-called edge effect (Lovejoy *et al.*, 1986). The opening of the forest allows both wind and light to penetrate (**589**) in a way that they do not in closed continuous forest. After only two years of isolation of small reserves (one hectare and 10 hectares), many dead and broken trees occurred around the edge and invasive species began to enter (Rankin de Merona *et al.*, 1994).

For plants, a disadvantage of a small reserve is often the loss of agents of pollination and seed dispersal. In the Manaus project, Powell and Powell (1987) demonstrated the loss of euglossine bees in the smaller fragments. These bees are important pollinators of orchids and also of many tree species, including *Bertholletia excelsa* (Brazil nut, **590, 591**) and many other species in the same family (Lecythidaceae). The loss of seed predators from an area would also affect the distribution of tree species (Terborgh, 1986, 1992). Brook *et al.* (2003) reported the rapid loss of species in Singapore, where the remaining forest has been reduced to extremely small areas (0.25% of the country, **592**). The rapid extinction of many species is happening in Singapore, and is predicted for much of the rest of southeast Asia. A significant problem with fragmentation of the forest is that it can lead to reduced gene flow and the increased possibility of inbreeding, especially for wind-pollinated species. For insect-pollinated species, even if there are connecting corridors, gene flow is likely to be much reduced relative to a continuous forest. There is little doubt that for optimum conservation of tropical rain forest trees, reserve areas large enough to maintain both biological interactions and adequate population size are essential.

591 A specimen of *Bertholletia excelsa* (Brazil nut tree) – the survival of this species is dependent on the pollination of its flowers (see **590**) by the carpenter bee (Photo copyright of R.E. Schultes.)

However, in some cases where the forest is not so species diverse, a network of smaller reserves could be better. Each reserve could be sited in a different habitat to preserve all available ecosystems. When this is done, the smaller reserves can be more effective than a single large one that does not have as much ecological variation. The SLOSS debate (single large or several small) will continue to rage, but most conservationists agree that the optimal number and size of reserves depend on the target species and the type of ecosystem to be preserved. Larger reserves are generally more effective for tree species unless the forest is very uniform. Small reserves can be more effective for annual herbaceous species, which often occur in dense localized stands (Lesica and Allendorf, 1992).

Godefroid and Koedam (2003) conducted an interesting study of the woodland patches within the city of Brussels, where one large woodland and several small areas remain. They found that the large wood still harboured species missing in the smaller one, and that 23 species of plants had a significantly higher frequency in this area. Six species had a higher frequency in the smaller woods but species of a higher conservation value, such as ancient forest species and rare species, had a higher frequency in the larger wood. The authors concluded that, although their overall data showed a higher conservation value for large woods, some plant functional groups (for example woodland species versus ancient forest species) responded differently to fragmentation. For establishing conservation strategies, studies considering the biotic characteristics of remnants should focus on the species number of particular functional groups, especially those with a high conservation value.

Small reserves can be rendered more effective when they are linked by vegetative corridors that allow the movement of animal pollinators and seed dispersers from one area to another. If gene flow can occur from one small area to another, then small reserves can be effective. Corridors by definition are usually narrow strips of land or even hedgerows. They therefore have a high edge-to-area ratio and are prone to edge effects. Nevertheless, they have often been shown to be effective. The restoration of degraded rain forest land for corridors is an important aspect of conservation (Lamb *et al.*, 1997).

POPULATION SIZE

The number of individuals needed to effectively conserve the genetic integrity of a species is a much debated issue (Soulé, 1987). This is often expressed as minimum viable population (MVP). Lawrence and Marshall (1997) suggested that an MVP of 5,000 is necessary, whereas Frankel and Soulé (1981) proposed from 500 to 2,000 individual plants, and Hawkes (1991) recommended 1,000. In an area of temperate forest where one or a few species dominate (585), it would only take a small area to preserve 5,000 individuals. However, to do this in just one place could be disastrous if it was hit by a storm, fire, or drought. Much more effective is a series of reserves, each with a genetically viable population, so that a back-up exists in case of catastrophe.

The structure of tropical rain forest is completely different from that of temperate region forests. In

592 In Singapore, the areas of tropical rain forest have been very drastically reduced, with the concomitant loss of biodiversity, due to the impact of humans.

Table 15 The frequency of species on a three-hectare rainforest plot on the Rio Xingu, Brazil (from Campbell *et al.*, 1986)

A. TEN MOST ABUNDANT SPECIES

Species	Number of individuals per three hectares
Cenostigma macrophyllum	184
Attalea phalerata	79
Theobroma speciosum	48
Neea altissima	45
Rinorea juruana	33
Lecythis retusa	32
Hirtella piresii	29
Inga sp.	29
Alexa imperatricis	28
Guatteria macrophylla	25

B. RARE SPECIES

Number of species with three individuals per three hectares	29
Number of species with two individuals per three hectares	54
Number of species with one individual per one hectare	125

C. TOTAL NUMBER OF SPECIES ON THREE-HECTARE PLOT

	265

rain forest there can be up to 300 different species on a single hectare (Gentry, 1988; Valencia *et al.*, 1994). The structure of the population is also different. Most inventories of areas of rain forest show that, in any hectare, a few species occur in reasonable numbers, but the majority are very sporadic. Data from an inventory of three hectares of rain forest in the Rio Xingu region of Pará State in Brazil (**584**) are given in *Table 15* (Campbell *et al.*, 1986). It would not take a reserve of many hectares in that region to preserve a population of 5,000 individuals of both *Cenostigma macrophyllum* and *Attalea phalerata*, which occur there in reasonable numbers (184 and 79 individuals per three hectares, respectively). However, most of the species represented on that sample plot have only one or two individuals in the three hectares (179 species out of a total of 265).

It would therefore take a very much larger area to include 5,000 individuals of any of these other species. This is a further argument for large reserve size when conserving tropical trees.

Recent conservation programmes for various rare species of plants have been successful in restoring populations of species that had been reduced to only a few individuals, for example *Trochetiopsis melanoxylon* (Saint Helena ebony, **593**) for which only two individuals existed, and *Hibiscadelphus woodii* from Kauai in Hawaii, with less than 10 individuals. When such drastic reduction occurs, the new population that has been through a genetic bottleneck will have a weaker genetic basis from which to resist pests and diseases. It is much better to maintain viable populations of 2,000–5,000 individuals.

CONSERVATION GENETICS

In recent years powerful new tools have been added to forest conservation through the use of molecular techniques. Methods such as gel electrophoresis or random amplified polymorphic DNA (RAPD) analysis of genetic variation enables the assessment of population structure. There is now a large variety of molecular marker types that are used for the analysis of genetic variation in plants. A summary of some of the methods is given by Newbury and Ford-Lloyd (1997). With the knowledge of the genetic structure of the population of a rare species, it is possible to prioritize which populations or individuals to conserve to achieve the greatest genetic diversity. For the conservation of a rare species it is necessary to maintain the gene pool and to conserve all the alleles in the population. Molecular techniques are used to define patterns of gene flow, and patterns and dynamics of genetic diversity, and thereby determine minimum viable population sizes.

THE HUMAN COMPONENT

The discussion about reserve and population size is necessary because forests and woodlands are being progressively destroyed by many different types of human activity. Thus, effective conservation has to be based on a balanced interaction with local human populations (**594, 595**). The most widely accepted large reserve design is that of the 'Man and the Biosphere' programme of UNESCO (Gadgil, 1983; Batisse, 1986; Cox, 1993). A biosphere reserve has a central core area which is totally protected, surrounded by a buffer zone where research, tourism, and educational activities are allowed and, in some cases, where indigenous peoples carry out their subsistence activities. In ideal situations this is surrounded by a transitional zone where such activities as sustainable extraction of timber, manipulation research, and ecosystem restoration take place.

The only activities that should be permitted in the core zone are monitoring by scientists, and vigilance patrols by reserve officials. The area needs to be large enough to protect viable populations of the species for which it was created. The existence of a buffer zone reduces edge effects and often extends the population size of many species. The success of

593 The flower of *Trochetiopsis melanoxylon* (Saint Helena ebony) – this is a highly endangered species and almost extinct in the wild, where only a few individuals have survived. (Photo copyright of M. Maunder.)

594 View of the Reserva Ecologica Guapi-açú (REGUA), part of the recently created Tres Picos Reserve of Rio de Janeiro State.

595 Group of Amazonian rubber tappers, who live by sustainable extraction of products from the forest. The establishment of extractivist reserves for them in Brazil is helping to conserve some parts of the Amazon forest.

biosphere reserves depends on establishing a good relationship with local people. The buffer zone is an area where the local people have access to traditionally used species, but they must also become involved in the protection of the core area. Many biosphere reserves have now been established around the world, and many other reserves try to follow the model as far as is possible in the local circumstances.

An example of an effective conservation programme involving the local people is the Mamirauá Reserve on the Rio Japurá in Amazonian Brazil. This Reserve protects a large area of the much-destroyed floodplain forest, as well as an extensive lake system of great importance for fish and aquatic mammals (**596, 597**). The success of this reserve is based on the involvement of the local inhabitants, who continue subsistence living within the reserve, while commercial fishing and timber activities are excluded. In Mamirauá the forest is well preserved and populations of manatee and the *Arapaima gigas* (giant piraracu fish) have increased remarkably, while they have become rare or extinct in much of the Amazon region.

In effect, the 1,000 families that reside in the reserve have become forest guards. This reserve has also been the location for much important scientific research on the floodplain forest and lakes of Amazonia. In this case, a very large complex of reserves has now been set up because Mamirauá has been linked to the Jaú Reserve in the Rio Negro region, through the establishment of the intervening area as a reserve. The key to the success has been the good relationship between the scientists involved and the local people. The reserve is also open to carefully planned and controlled ecotourism.

There are many other types of effective reserves besides the biosphere ones. These include managed resource areas, national parks (**598**), protected watersheds, strictly protected local reserves, and even sacred forested groves in such places as India and Thailand. The International Union for Conservation of Nature (IUCN) lists six categories of protected reserve (*Table 16*). Any of these categories may be used to effectively preserve forest and woodland. Generally, both smaller areas and areas from which products are being extracted will require more active management.

A reserve of the much depleted Atlantic coastal rain forest of Brazil that has recently been established is the Reserva Ecologica Guapi-açú (REGUA). This has been achieved by raising funds in the UK for land purchase of a central area of forest (**594**), which is administered by a Brazilian foundation. The effectiveness of the reserve has been greatly increased by its being adjacent to a watershed area that is owned by a brewery. In order to preserve water quality the brewery found it necessary to purchase a whole watershed that is covered by rain forest. This considerably increases the significance of REGUA, and the REGUA forest wardens are allowed to patrol the neighbouring watershed as part of the programme to stop hunting in the area. Many reserves exist to maintain a broader ecological function such as watershed or climate, and these also

596 Spreading *Ficus* (fig) tree in the floodplain of the Mamirauá Reserve on the Rio Japurá in Amazonian Brazil.

597 Floodplain várzea forest of the Mamirauá Reserve in Amazonian Brazil.

Table 16 IUCN categories for protected areas

Category 1. **Strict Protection.** Nature reserves and wilderness areas that are strictly managed for nature conservation and science where human intervention and exploitation are not allowed.

Category 2. **Ecosystem Conservation and Tourism.** Areas such as national parks where visitors are allowed for recreation and inspiration without harming the area's ecosystem.

Category 3. **Conservation of Natural Features.** Areas that preserve a single natural feature or historic site and incidentally conserve forest and biodiversity.

Category 4. **Conservation through Active Management.** Areas with more specific active management in order to maintain a specific habitat or species.

Category 5. **Landscape/Seascape Conservation and Recreation.** Areas where the landscape rather than the biology is the primary focus.

Category 6. **Sustainable Use of Natural Ecosystems.** Protected areas that are managed for the sustainable use of natural products such as timber and non-timber forest products. The extractivist reserves of Brazil world fall in this category.

provide effective biological conservation.

It will often only be possible to maintain a particular reserve when it helps to cover its costs through sustainable production of timber or non-timber forest products (**595, 599**), or by ecotourism. The involvement of any of these activities needs to be carefully managed if conservation goals are to be achieved. In several places this may have come about through the action of local residents. In the states of Acre and Amapá in Brazil, a number of extractivist reserves have been established largely through the activism of local rubber tappers (**595**) and Brazil nut gatherers. These reserves, where extraction of non-timber products is allowed, are largely managed by the local communities that depend upon the forest for their existence (Nepstad and Schwartzman, 1992; Wilke, 1994; Peters and Hammond, 1990).

In some areas the extraction of non-timber products is combined with timber extraction. For example, in the Sinharaja Forest Reserve (Sri Lanka), several products are gathered, including *Calamus*

598 Well-preserved rain forest of Brunei. The income from oil in that country has deflected the need to cut the rain forest.

599 A charcoal burner in the forest reserve of Wakehurst Palace in West Sussex, England.

(rattan cane species), *Caryota urens* (kithul or toddy palm, **600**, **601**) for the production of a sugar from the sap (jaggery), wild cardamom, and the medicinal herb *Coscinium fenestratum*. All of these operations were performed better in selectively logged-over forest than in undisturbed forest (Gunatilleke *et al.*, 1995). There are also many examples of the over-harvesting of non-timber products, for example rattans in southeast Asia (Dransfield, 1989) and *Mauritia flexuosa* (aguaje palm, **602**) in Peru (Vasquez and Gentry, 1989). In the latter case, this was entirely unnecessary because many female trees were being felled just to gather a one-time harvest of the fruit.

Many indigenous peoples of the forest have maintained the forest and the biodiversity of the areas in which they live. Some of the indigenous reserves of the Amazon have much intact forest,

600 Specimen of *Caryota urens* (toddy palm) growing in Sri Lanka.

601 *Caryota urens* (toddy palm). As its common name suggests, the palm sap contains about 10% sucrose which, when fermented, produces a popular alcoholic drink. (Photo copyright of Bryan Bowes.)

602 Specimen of *Mauritia flexuosa* (aguaje palm) growing in Peru – the fruit from female trees was gathered unsustainably because the trees were felled to obtain a one-off harvest.

603 Chácobo Indian preparing cassava flour. (Photo copyright of B.M. Boom.)

although often the game species have been depleted. Studies have shown the large extent to which these people rely on the forest for their existence (**603**; Balée, 1987; Boom, 1987; Prance *et al.*, 1987). The Chácobo Indians of Bolivia used 82% of the tree species of an area of forest sampled by Boom (1987). When the forest is so important as a source of their livelihood, the Indians are not so likely to remove all its resources.

This section has shown that in order to achieve effective conservation of forest and woodland, the human element must be taken into consideration. Fully protected biological reserves and core areas are important, but much biodiversity will not be saved unless sustainable use is also considered. This will enable the ultimate conservation of much larger areas, and also makes a much more convincing argument politically. Sustainable management is that which both conserves biodiversity, and provides long-term goods and services.

MAINTAINING BIOLOGICAL INTERACTIONS

An essential for any conservation area is the need to maintain the web of interaction between species. This generally becomes harder in smaller areas, where some animal species tend to drop out.

The ecosystem will not survive if pollination, seed dispersal, and other biological interactions cease. For example, the flowers of the *Crescentia cujete* (Amazonian calabash) are pollinated by bats. When this happens the young fruit begin to develop and quickly produce extra-floral nectaries on the exocarp (Elias and Prance, 1978). These nectaries are visited by aggressive ants which afford protection to the fruit (**604**) during this early stage of development. As the fruit enlarges it becomes woody and no longer needs its insect protectors, and the nectaries disappear. This most useful indigenous utensil of the Amazon region depends on bats and ants for its continued propagation. Another crucial interaction is between the roots of trees and mycorrhizal fungi in both temperate and tropical forests. In restoration projects it is particularly important to recuperate the other organisms upon which the trees depend for their existence.

In temperate forests many of the tree species, such as pines and oaks, are wind-pollinated and individuals are close to other individuals of their species. In tropical rain forest many species are rare and scattered. However, there is a high incidence of outcrossing in such species (**605**; O'Malley and Bawa, 1987; Murawski, 1995), and this depends on specialized pollinators such as

604 Protective ants on the young fruit of a calabash tree (*Crescentia cujete*) in Brazil.

605 *Crateva benthamii*, a typical riverside species (which is bat-pollinated) of the floodplain forest at the Mamirauá Reserve in Amazonian Brazil.

bees (**590**), bats, or birds. Forest conservation involves the conservation of the whole ecosystem (**584, 598**).

In Finland, the reserves that have been set up to preserve the genetic diversity of the country's major commercial forest species are thought of as 'functional pollination units' (Koski, 1991). This is a network of gene reserve forests to maintain the full genetic diversity of these species, through the maintenance of the breeding systems.

RESERVE MANAGEMENT

To be effective a conservation area must have a management plan which takes into consideration many of the factors already mentioned above, such as population sizes and the maintenance of biological interactions. Since management plans are generally for the use of field workers who are often local people, they need to be simple and in the local language without an excess of technical jargon. Management plans must be based on sound research data about species composition, geographic information, soils, climate, etc. The plan itself needs to include this basic information as well as maps of the boundaries, the history of the area, and the ownership of the land. Such a plan is at present underway in the Yaboti Biosphere Reserve of Misiones Province in Argentina.

A plan should also include a summary of the research data, such as biological inventories and soil surveys; it needs to state clearly who is permitted access to the different parts of the reserve and for what activities. It should outline the individual roles and targets for all those involved in management, from the director to the forest rangers and maintenance staff. The plan must also include such items as the budget for a set period and how this will be obtained, and details of a programme for monitoring the effectiveness of the reserve, as well as the use of the reserve for education. In many cases it is also necessary to assess the risks to the integrity, or mission, of a reserve, and to consider contingency plans to deal with these situations. Many reserves are ineffective in their conservation goal because they lack good planning and fail to look ahead.

EDUCATION AND CAPACITY BUILDING

One of the most important ways in which to maintain interest in and support for conservation areas, whether they be small wood lots in an urban area or large tracts of rain forest, is through their use for education. Many reserves around the world now have active education programmes at many different levels, from primary-school children to doctoral research students. Reserves are where the conservationists of the future are inspired and trained.

In the REGUA reserve of Brazil, an active programme among school children is even getting the message over to their parents, who hunt illegally in the reserve. Also in Brazil, in Pará state, a team of scientists has incorporated their work on the productivity of local fruit trees into a popular, abundantly illustrated book for semi-literate local people. The book presents, in simple terms, the long-term advantage of preserving the fruit trees that provide marketable fruit rather than selling them off for timber for a small, one-time profit (Pye-Smith, 2003). This education programme with these people has certainly led to the conservation of the forests from where fruits and other products are extracted.

Many forests are both valued and maintained because of their use in education, and many conservationists have been trained through the education and research programmes of reserves.

CONCLUDING REMARKS

Although the officially protected areas around the world are said to cover 8.8% of the earth's land surface, this is a gross overestimate because many reserves in these statistics have no protection at all and lack any plans for their management. A survey of reserves in ten developing countries (Dudley and Stolton, 1999) found that only 1% of those reserves could be considered secure, and more than 20% were suffering from degradation. In many places, such as Indonesia and Brazil, illegal logging is carried out in reserves. Many other reserves are threatened by pollution and by climate change. Another frequent threat is from introduced alien species. *Acacia* and *Melaleuca* trees from Australia have taken over large areas of the Cape Province of South Africa, while Hawaii contains almost as many foreign species as its native flora, with *Mikania*, *Vernonia*, and *Melastoma* (**606**) particularly intrusive.

The conservation of forest and woodland is vital for the survival of many species of plants, animals, and fungi. It is in the forests of the world that much carbon is stored, and forests are also vital for the preservation of climate and water quality (see also Chapter 10). To avoid the worst effects of climate change, we need more, not less, forest and woodland. Forests everywhere in the world store carbon, and so the need for conservation and restoration is universal in any region that is ecologically appropriate for forest growth. Our future depends on trees, and so their conservation is absolutely vital.

606 *Melastoma tomentosa*, one of the many invasive species in Hawaii that are destroying the native forest.

References and general reading

CHAPTER 1 SURVEY OF TREES: THEIR GLOBAL SIGNIFICANCE, ARCHITECTURE, AND EARLY EVOLUTION

REFERENCES

Anonymous (2003). Oldest living tree to be cloned. *Plant Talk*, **31**: 16.

Anonymous (2004). El Grande. *Tree News*, Spring/ Summer 2004. The Tree Council, London, p. 10.

Anonymous (2005). *Tree Aid Update*, Summer 2005. Tree Aid, Bristol.

Bell A D (1991). *Plant Form. An Illustrated Guide to Flowering Plant Morphology*, Oxford University Press, Oxford.

Dix N J and Webster J (1995). *Functional Ecology*, Chapman & Hall, London.

Esau K (1965). *Plant Anatomy*, John Wiley, New York.

Foster A S and Gifford E M (1974). *Comparative Morphology of Vascular Plants*, 2nd edn, W H Freeman, San Fransisco.

Jane F W (1970). *The Structure of Wood*, 2nd edn, Adam & Charles Black, London.

Lamb R (2002). The changing face of the world's forests. *Tree News*, Autumn/Winter 2002. The Tree Council, London, pp. 24–27.

Lewington A (2003). *Plants for People*, Transworld Publishers, London.

Lewington A and Parker E (1999). *Ancient Trees – Trees that Live for a Thousand Years*, Collins and Brown, London.

Longman K A and Jenik J (1987). *Tropical Forest and its Environment*, 2nd edn, Longman, Harlow.

Oldfield S, Lusty C, and MacKinven A (1998). *The World List of Threatened Trees*, World Conservation Press, Cambridge.

Packham J R, Harding D J L, Hilton G M, and Stuttard R A (1992). *Functional Ecology of Woodlands and Forests*, Chapman & Hall, London.

Rackham O (2004). Trees underground. *Tree News Sylva*, Autumn/Winter 2004. The Tree Council, London.

Thomas P (2000). *Trees: Their Natural History*, Cambridge University Press, Cambridge.

Tomlinson P B (2003). Palms in botanic gardens. *Plant Talk*, **34**, 46–47. National Tropical Botanical Garden, Hawaii.

Willis K J and McElwain J C (2002). *The Evolution of Plants*, Oxford University Press, Oxford.

GENERAL READING

Akeroyd J (2000). Tapping into a reverence for trees. *Plant Talk*, **22/23**: 53–56.

Bowes B G and Mauseth J D (2008). *Plant Structure – A Colour Guide*, 2nd edn, Manson Publishing, London.

Bramwell D (2002). How many plant species are there? *Plant Talk*, **28**: 32–34.

Ingrouille M (1992). *Diversity and Evolution of Land Plants*, Chapman & Hall, London.

Langenheim J H (2003). *Plant Resins: Chemistry, Evolution, Ecology and Ethnobotany*, Timber Press, Portland.

Mattheck C and Breloer H (1994). *The Body Language of Trees: A Handbook of Failure Analysis*, HMSO, London.

May J (2004). Cork in crisis. *Tree News*, Spring/Summer 2004. The Tree Council, London, pp. 17–28.

Pakenham T (1996). *Meetings with Remarkable Trees*, Weidenfeld and Nicolson, London.

Sporne K R (1974). *The Morphology of Gymnosperms*, 2nd edn, Hutchinson, London.

Sporne K R (1974). *The Morphology of Angiosperms*, Hutchinson, London.

CHAPTER 2 NORTHERN BOREAL AND MONTANE CONIFEROUS FORESTS
GENERAL READING

Burns R M and Honkala B H (eds) (1990). *Silvics of North America, Volume 1, Conifers.* Agriculture Handbook 654, Forest Service, United States Department of Agriculture, Washington DC.

Farjon A (1990). *Pinaceae. Drawings and descriptions of the genera* Abies, Cedrus, Pseudolarix, Keteleeria, Nothotsuga, Tsuga, Cathaya, Pseudotsuga, Larix *and* Picea. [Regnum Vegetabile Vol. 121]. Koeltz Scientific Books, Königstein.

Farjon A (2001). *World Checklist and Bibliography of Conifers,* 2nd edn, Royal Botanic Gardens, Kew, England.

Farjon A (2004). *Pines. Drawings and Descriptions of the Genus* Pinus, 2nd edn, E J Brill Academic Publishers, Leiden.

Farjon A (2005). *A Monograph of Cupressaceae and* Sciadopitys. Royal Botanic Gardens, Kew, England.

Farjon A and Page C N (2000). An action plan for the world's conifers. *Plant Talk*, **21/23**: 43–47.

Lanner R M (1999). *Conifers of California*, Cachuma Press, Los Olivos.

Richardson D M (ed.) (1998). *Ecology and Biogeography of* Pinus, Cambridge University Press, Cambridge.

CHAPTER 3 TEMPERATE DECIDUOUS AND TEMPERATE RAIN FORESTS
REFERENCES

Alaback P (1989). Logging of Temperate Coastal Rain Forests, Ecological Factors to Consider. *Proceedings of the 1989 'Watershed Symposium'*, United States Forest Service.

Alaback P (1990). *Comparative Ecology of Temperate Rain Forests of the Americas along Analogous Climatic Gradients.* United States Forest Service, Pacific Northwest Research Station.

Franklin J and Waring R (1980). Distinctive Features of the Northwestern Coniferous Forest: Development Structure and Function. *Proceedings of the 40th Annual Biology Colloquium*, Oregon State University Press.

Juneau A K and Fujimori T (1971). Primary productivity of a young *Tsuga heterophylla* stand and some speculations about biomass of forest communities on the Oregon Coast. United States Department of Agriculture Forest Research Paper, Washington DC.

Weigand J (1990). Coastal temperate rain forests: Definition and global distribution with particular emphasis on North America. Unpublished report prepared for Ecotrust/Conservation International.

GENERAL READING

Coombes A J (2004). *An Illustrated Directory of Trees and Shrubs*, Salamander Books, London.

Echeverria C (2002). Ecological Disaster in Southern Chile. *Plant Talk*, **28**: 14. National Tropical Botanic Garden, Hawaii.

Foster B (1989). The West Coast: Canada's rain forest. In: Hummel and Monte (eds), *Endangered Spaces*, Key Porter Books, Toronto, pp. 196–210.

Hoffmann A E (1994). *Flora Silvestre de Chile zona Araucana Arboles, Arbustos y Enredaderas Lenosas*, 3rd edn, Ediciones Fundación Claudio Gay, Santiago de Chile.

Holliday I (1989). *Australian Trees*, Hamyln, Australia.

Janik J (1979). *Pictorial Encyclopedia of Forest*, Hamlyn, Feltham, Middlesex.

Johnston V R (1994). *California Forests and Woodlands A Natural History*, University of California Press, Berkeley.

Little E L (1994). *National Audubon Society Field Guide to North American Trees Western Region*, Knopf, New York.

Marinelli J (2004). *Plant*, Dorling Kindersley, London.

Mitchell A (1974). *A Field Guide to the Trees of Britain and Northern Europe*, Collins, London.

Packham J R, Harding D J L, Hilton G M, and Stuttard R A (1992). *Functional Ecology of Woodlands and Forests*, Chapman & Hall, London.

Pavlik B M, Muick P C, Johnson S, and Popper M (1991). *Oaks of California*, Cachuma Press, Los Olivos.

Phillips R (1978) *Trees in Britain, Europe and North America*, Macmillan, London.

Rohrig E and Ulrich B (eds) (1991) *Ecosystems of the World: 7, Temperate Deciduous Forests*, Elsevier, Amsterdam.

Rottmann J (1988). *Bosques de Chile Chile's Woodlands*, 2nd edn, IGES, Santiago de Chile.

Russell T and Cutler C (2004). *Trees an Illustrated Identifier and Encyclopedia*, Anness Publishing, London.

Salmon J T (1986). *The Reed Field Guide to New Zealand Trees*, Reed, Auckland.

CHAPTER 4 TEMPERATE MIXED EVERGREEN FORESTS

REFERENCES

Beard J S (1990). *Plant Life of Western Australia*, Kangaroo Press, Sydney.

Braun-Blanquet J (1936). La chênaie d'Yeuse languedocienne (*Quercion ilicis*). *Memoires de la Societé d' Études des Sciences Naturelles de Nîmes*, **5**: 1–147.

Brooker M I H and Kleinig D A (2001). *Field Guide to Eucalypts, Vol. 2, South-western and Southern Australia*, 2nd edn, Blooming Books, Melbourne.

Brooker M I H and Hopper S D (2002). Taxonomy of species deriving from the publication of *Eucalyptus* subseries *Cornutae* (Myrtaceae). *Nuytsia*, **14**(3): 325–360.

Cornett J W (1999). *The Joshua Tree*. Nature Trails Press, Palm Springs.

Cowling R M, Rundel P W, Lamont B B, Arroyo M K, and Arianoutsou M (1996). Plant diversity in Mediterranean-climate regions. *Trends in Ecology and Evolution*, **11**: 362–366.

Cowling R M, Richardson D M, and Pierce S M (eds) (1997). *Vegetation of Southern Africa*. Cambridge University Press, Cambridge.

Dallman P R (1998). *Plant Life in the World's Mediterranean Climates*. California Native Plant Society, University of California Press, Berkeley.

Dawson J and Lucas R (2000). *Nature Guide to the New Zealand Forest*. Godwit, Auckland.

Di Castri F, Goodall D W, and Specht R L (1981). *Mediterranean Type Shrublands*. Elsevier, Amsterdam.

Goldblatt P (1993). *The Woody Iridaceae* Nivenia, Klattia *and* Witsenia *Systematics Biology and Evolution*. Timber Press, Oregon.

Goldblatt P and Manning J (2000). Cape plants a conspectus of the Cape flora of South Africa. *Strelitzia*, **9**: 1–743.

Hoffmann A E (1994). *Flora Silvestre de Chile zona Araucana Arboles, Arbustos y Enredaderas Lenosas*, 3rd edn, Ediciones Fundacion Claudio Gay, Santiago de Chile.

Hoffmann A E (1995). *Flora Silvestre de Chile zona Central*, 3rd edn, Ediciones Fundacion Claudio Gay, Santiago de Chile.

Hopper S D and Gioia P (2004). The southwest Australian floristic region: evolution and conservation of a global biodiversity hotspot. *Annual Review of Ecology, Evolution and Systematics*, in press.

Johnston V R (1994). *California Forests and Woodlands A Natural History*, University of California Press, Berkeley.

Kirkpatrick J B and Backhouse S (1985). *Native Trees of Tasmania*, Pandani Press, West Hobart.

Lanner R M (1999). *Conifers of California*, Cachuma Press, Los Olivos.

Maslin B R *et al.* (2001). *Acacia. Flora of Australia Vols 11A and 11B*, Australian Biological Resources Study, Canberra, and CSIRO, Melbourne.

Noss R F (ed.) (2000). *The Redwood Forest History, Ecology and Conservation of the Coast Redwoods*, Island Press, Washington DC.

Ornduff R, Faber P M, and Keeler-Wolf T (2003). *Introduction to California Plant Life*, 2nd edn, University of California Press, Berkeley.

Pauw A and Johnson S (1999). *Table Mountain A Natural History*, Fernwood Press, Vlaeberg.

Pavlik B M, Muick P C, Johnson S, and Popper M (1991). *Oaks of California*, Cachuma Press, Los Olivos.

Pignatti S (1998). *I boschi d'Italia. Sinecologia e biodiversità*, UTET, Torino.

Pignatti Wikus E, Pignatti G, and Hopper S D (2001). Mallee communities along roadsides in south-western Australia. *Studia Geobotanica*, 20: 3–16.

Playford P (1998). *Voyage of Discovery to Terra Australis by Willem de Vlamingh in 1696–97*. Western Australian Museum, Perth.

Quézel P and Medail F (2003). *Ecologie et biogeograpihie des forêts du bassin méditerranéen*. Elsevier, Amsterdam.

Rottmann J (1988). *Bosques de Chile Chile's Woodlands*, 2nd edn, IGES, Santiago de Chile.

Salmon J T (1986). *The Reed Field Guide to New Zealand Trees*, Reed, Auckland.

Steane D A, Nicolle D, MacKinnon G E, Vaillancourt R E, and Potts B M (2002). Higher-level relationships among the eucalypts are resolved by ITS-sequence data. *Aust. Syst. Bot.*, 15: 49–62.

van der Merwe I (1998). *The Knysna and Tsitsikamma forests Their History, Ecology and Management*, Tafelberg, Cape Town.

Zegers C D (1995). *Arboles Nativos de Chile Chilean Trees*, Marisa Cuneo Ediciones, Valdivia.

CHAPTER 5 TROPICAL AND SUB-TROPICAL RAIN AND DRY FORESTS

REFERENCES

Absy M L, Prance G T and Barbosa E M (1988). Inventário floristoco de floresta natural na área da estrada Cuiabá-Porto Velho (BR364). *Acta Amazonica*, 16/17 (Suppl):85–121.

Almeida S S, Lisboa P L B, and Silva A S L (1993). Diversidade florística de uma comunidade arbórea na estação científica "Ferreira Penna", em Caxiuanã (Pará). *Bol. Mus. Paraense Emílio Goeldi, Sér. Bot.*, 9(1): 93–128.

Ashton P S (1989). Dipterocarp reproductive biology. In: Lieth H and Werger M J A (eds), *Tropical Rain Forest Ecosystems. Ecosytems of the World 14B*, Elsevier, Amsterdam, pp. 219–240.

Balslev H, Luteyn J, Øllgaard B, and Holm-Nielsen L B (1987). Composition and structure of adjacent unflooded and floodplain forest in Amazonian Ecuador. *Opera Botanica*, 92: 37–57.

Boom B M (1986). A forest inventory in Amazonian Bolivia. *Biotropica*, 18: 287–294.

Cain S A, Castro G M O, Pires J M, and Silva N T (1956). Application of some phytosociological techniques to Brazilian rain forest. *Amer. Jour. Bot.*, 43: 911–941.

Campbell D G, Daly D C, Prance G T, and Maciel U N (1986). Quantitative ecological inventory of terra firme and várzea tropical forest in the Rio Xingu, Brazilian Amazon. *Brittonia*, 38: 369–393.

Gentry A H (1988). Tree species richness of upper Amazonian forests. *Proc. Natl. Acad. Sci. US*, 85: 156–159.

Groombridge B and Jenkins M B (2002). *World Atlas of Biodiversity*, University of California Press, Berkeley.

Grubb P J and Whitmore T C (1966). A comparison of montane and lowland rain forest in Ecuador II. The climate and its effects on the distribution and physiognomy of the forests. *Jour. Ecol.*, 54: 303–333.

Hamilton A (1989). African forests. In: Lieth H and Werger M J A (eds), *Tropical Rain Forest Ecosystems. Ecosystems of the World 14B*, Elsevier, Amsterdam, pp. 155–182.

Kadavul K and Parthasarathy N (1999). Structure and composition of woody species in tropical semi-evergreen forest of Kalrayan hills, Eastern Ghats, India. *Trop. Ecology*, 40(2): 247–260.

Lisboa P L B (1989). Estudo florístico da vegetação arbórea de uma floresta secundária em Rondônia. *Bol. Mus. Paraense Emílio Goeldi, Sér. Bot.*, 5(2): 145–162.

Milliken W, Miller R P, Pollard S R, and Wandelli E V (1992). *Ethnobotany of the Waimiri Atroari Indians of Brazil*. Royal Botanic Gardens, Kew.

Mori S A, Rabelo B V, Tsou C–H, and Daly D (1989). Composition and structure of an eastern Amazonian forest at Camaipi, Amapá, Brazil. *Bol. Mus. Paraense, Emílio Goeldi, Sér. Bot.*, **5**(1): 3–18.

Pires J M and Prance G T (1985). The vegetation types of the Brazilian Amazon. In: Prance G T and Lovejoy T L (eds), *Key Environments: Amazonia*, Pergamon Press, Oxford, pp. 109–145.

Prance G T (1979). Notes on the vegetation of Amazonia III. The terminology of Amazonian forest types subject to inundation. *Brittonia*, **31**: 26–38.

Prance G T (1989). American tropical forests. In: Lieth H and Werger M J A (eds), *Tropical Rain Forest Ecosystems. Ecosytems of the World 14B*, Elsevier, Amsterdam, pp. 99–132.

Proctor J, Anderson J M, Chai P, and Vallack H W. (1983). Ecological studies in four contrasting lowland rainforests in Gunung Mulu National Park, Sarawak 1. Forest environment, structure and floristics. *Journal of Ecology*, **71**: 237–260.

Ramanujam M P and Kadamban D (2001). Plant diversity of two tropical dry evergreen forests in the Pondicherry region of South India and the role of belief systems in their conservation. *Biodiversity and Conservation*, **10**:1203–1217.

Reitsma J M (1988). Végétation forestière du Gabon. *Technical Series Tropenbos*, **1**: 1–142.

Silva A S L da, Lisboa P L B, and Maciel U N (1992). Diversidade florística e estrutura em floresta densa da bacia do Rio Juruá-Am. *Bol. Mus. Paraense Emílio Goeldi, Sér. Bot.*, **8**(2): 203–258.

Spichiger R, Loizeau P-A, Latour C and Barriera G (1996). Tree species richness of a South-Western Amazonian forest (Jenaro Herrera, Peru, 73°40'W; 4°54'S). *Candollea*, **51**: 559–577.

Thomas W W, and Carvalho A M de (1997). Atlantic moist forest of Southern Bahia. In: Davis S D, Heywood V H, MacBryde O H, and Hamilton A C (eds), *Centres of Plant Diversity: A guide and Strategy for their Conservation*, IUCN-WWF, London, pp. 364–368.

Vásquez-Martínez R and Phillips O L (2000). Allpahuayo: Floristics, structure and dynamics of a hyperdiverse forest in Amazonian Peru. *Ann. Missouri Bot. Gard.*, **87**: 499–527.

Whitmore T C, Sidiyasa K, and Whitmore T J (1987). Tree species enumeration of 0.5 hectares on Halmahera. *Gardens Bull.*, **40**: 31–34.

Wright D, Jessen J H, Burke P, and Silva Garza H G de (1997). Tree and liana enumeration and diversity on a one hectare plot in Papua New Guinea. *Biotropica*, **29**: 250–260.

CHAPTER 6 WOODY THICKENING IN TREES AND SHRUBS
REFERENCES

Bell A D (1991). *Plant Form. An Illustrated Guide to Flowering Plant Morphology*, Oxford University Press, Oxford.

Bowes B G (1999). Wound-healing and vegetative regeneration of trees in the West of Scotland. *Glasgow Naturalist,* **23**(part 4): 12–18. Glasgow Natural History Society.

Longman K A and Jenik J (1987). *Tropical Forest and its Environment*, 2nd edn, Longman, Harlow.

Mattheck C and Breloer H (1994). *The Body Language of Trees: A Handbook of Failure Analysis*, HMSO, London.

Mauseth J D (2003). *Botany an Introduction to Plant Biology*, Jones and Bartlett, Boston.

Milner M E (1932). Natural grafting in *Hedera helix*. *New Phytologist*, **31**: 2–25.

Sharples S and Gunnery H (1933). Callus formation in *Hibiscus rosa-sinensis* and *Hevea brasiliensis*. *Annals of Botany*, **47**: 827–840.

Smith D M (1986). *The Practice of Silviculture*, John Wiley, New York.

Thomas P (2000). Trees: *Their Natural History*, Cambridge University Press, Cambridge.

Tomlinson P B (2003). Palms in botanic gardens. *Plant Talk*, **34**: 46–47. National Tropical Botanical Garden, Hawaii.

GENERAL READING

Bowes B G and Mauseth J D (2008). *Plant Structure – A Colour Guide*, 2nd edn, Manson Publishing, London.

Esau K (1965). *Plant Anatomy*, 2nd edn, Wiley, New York.

Foster A S and Gifford E M (1974). *Comparative Morphology of Vascular Plants*, 2nd edn, W H Freeman, San Fransisco.

Ingrouille M (1992). *Diversity and Evolution of Land Plants*, Chapman & Hall, London.

Jane F W (1970). *The Structure of Wood*, 2nd edn, Adam & Charles Black, London.

Langenheim J H (2003). *Plant Resins: Chemistry, Evolution, Ecology and Ethnobotany*, Timber Press, Portland.

Mauseth J D (1988). *Plant Anatomy*, Benjamin/Cummings, California.

May J (2004). Cork in crisis. *Tree News*, Spring/Summer 2004. The Tree Council, London, pp. 17–28.

Sporne K R (1974). *The Morphology of Gymnosperms*, 2nd edn, Hutchinson, London.

CHAPTER 7 THE ROLE OF CELL WALL POLYMERS IN DISEASE RESISTANCE IN WOODY PLANTS

REFERENCES

Brunner A M, Busov V B, and Strauss S H (2004). Poplar genome sequence: functional genomics in an ecologically dominant plant species. *Trends in Plant Science*, 9: 49–56.

De Ascenao A R D C F and Dubery I A (2000). Panama disease: cell wall reinforcement in banana roots in response to elicitors from *Fusarium oxysporum* f. sp. *cubense* race four. *Phytopathology*, 90: 1173–1180.

Smith C M, Rodriguez-Buey M, Karlsson J, and Campbell M M (2004). The response of the poplar transcriptome to wounding and subsequent infection by a viral pathogen. *New Phytologist*, 164: 123–126.

Valette C, Nicole M, Sarah J L, Boisseau M, Boher B, Fargette M, and Geiger J P (1997). Ultrastructure and cytochemistry of interactions between banana and the nematode *Radopholus similes. Fundamental and Applied Nematology*, 20: 65–77.

Vorwerk S, Somerville S, and Somerville C (2004). The role of plant cell wall polysaccharide composition in disease resistance. *Trends in Plant Science*, 9: 203–209.

GENERAL READING

Agrios G N (2005). *Plant Pathology*, 5th edn, Elsevier Academic Press, Burlington, MA.

Brett C T and Waldron K W (1996). *Biochemistry and Physiology of Plant Cell Walls*, Chapman & Hall, London.

Chilosi G and Magro P (1998). Pectolytic enzymes produced *in vitro* and during colonisation of melon tissues by *Didymella bryoniae*. *Plant Pathology*, 47: 700–705.

Mendgen K, Hahn M, and Deising H (1996). Morphogenesis and mechanisms of penetration by plant pathogenic fungi. *Annual Review of Phytopathology*, 34: 367–386.

Pearce R B (1996). Antimicrobial defences in the wood of living trees. *New Phytologist*, 132: 203–233.

Silva M C, Nicole M, Rijo L, Geiger J P, and Rodrigues C J (1999). Cytochemical aspects of the plant – rust fungus interface during the compatible interaction *Coffea arabica* (cv Caturra) *Hemileia vastatrix*. *International Journal of Plant Sciences*, 160: 79–91.

CHAPTER 8 MICROBIAL AND VIRAL PATHOGENS, AND PLANT PARASITES OF PLANTATION AND FOREST TREES

GENERAL READING

Butin H (1995). *Tree Diseases and Disorders: Causes, Biology, and Control in Forest and Amenity Trees*, Oxford University Press, Oxford.

Cooper J I (1993). *Virus Diseases of Trees and Shrubs*, 2nd edn, Chapman & Hall, London.

Fox R T V (2000). Armillaria *Root Rot: Biology and Control of Honey Fungus*, Intercept, Andover.

Hansen E M and Lewis K J (1997). *Compendium of Conifer Diseases*, APS Press, St. Paul.

Hawksworth F G and Wiens D (1996). *Dwarf Mistletoes: Biology, Pathology and Systematics*, USDA Forest Service Handbook 709, Washington DC.

Manion P D (1991). *Tree Disease Concepts*, 2nd edn, Prentice-Hall.

Pei M H and McCracken E R (2005). *Rust Diseases of Willow and Poplar*, CAB International, Wallingford.

Phillips D H and Burdekin D A (1992). *Diseases of Forest and Ornamental Trees*, 2nd edn, Macmillan, London.

Redfern D B and Gregory S C (1998). *Disease Diagnosis in Forest Trees*, HMSO, London.

Rizzo D M and Garbellotto M (2003). Sudden oak death: endangering California and Oregon forest ecosystems. *Frontiers in Ecology and the Environment*, **1**(4): 197–204.

Shaw C G and Kile G A (1991). Armillaria *Root Disease*. USDA Forest Service, Agriculture Handbook No. 691, Washington DC.

Sinclair W A and Lyon H H (2005). *Diseases of Trees and Shrubs*, 2nd edn, Cornell University Press, New York.

Strouts R G and Winter T G (1994). *Diagnosis of Ill-Health in Trees*, HMSO, London.

Tainter F H and Baker F A (1996). *Principles of Forest Pathology*, John Wiley & Sons, New York.

Watson D M (2001). Mistletoe – A keystone resource in forests and woodlands worldwide. *Annual Reviews of Ecology and Systematics*, **32**: 219–249.

Woodward S, Stenlid J, Karjalainen R, and Hüttermann A (1998). Heterobasidion annosum: *Biology, Ecology, Impact and Control*, CAB International, Wallingford.

CHAPTER 9 INSECT PESTS OF SOME IMPORTANT FOREST TREES
REFERENCES

Bain J (1976). Phoracantha semipunctata *(Fabricius) (Coleoptera: Cerambycidae)*. New Zealand Forest Sevice, Forest and Timber Insects in New Zealand, no 4.

Bakke A and Strand L (1981). Pheromones and traps as part of an integrated control of the spruce bark beetle *Ips typographus* some results from a control program in Norway in 1979 and 1980. *Nisk Norsk Institutt Skogforsk Rapport*, 0(5): 5–39.

Baumgras J E, Sendak P E, and Sonderman D L (1999). Ring shake in eastern hemlock: frequency and relationship to tree attributes. In: McManus K A, Shields K S, and Souto D R (eds), *Proceedings of the Symposium on Sustainable Management of Hemlock Ecosystems in Eastern North America*, USDA Forest Service General Technical Report NE-267, pp. 156–160.

Byers J A (1989). Behavioral mechanisms involved in reducing competition in bark beetles. *Holarctic Ecology*, **12**: 466–476.

Cabrera B J, Su N-Y, Scheffrahn R H, Oi F M, and Koehler P G (2001). *Formosan Subterranean Termite 1*. Entomology and Nematology Department, Florida Cooperative Extension Service, Institute of Food and Agricultural Sciences, University of Florida ENY-216. http://edis.ifas.ufl.edu/MG064.

Cerezke H F (1991). *Spruce Budworm*, Forestry Canada Northwest Region Northern Forestry Centre, Edmonton, Alberta, Forestry Leaflet 9.

Cheah C A S-J, Montgomery M S, Salom S M, Parker B, Skinner M, and Reardon R (2004). *Biological Control of the Hemlock Woolly Adelgid*, USDA Forest Service FHTET-2004-04.

Day K R, Halldórsson G, Harding S, and Straw N A (eds) (1998). *The Green Spruce Aphid in Western Europe: Ecology, Status, Impacts and Prospects for Management*, Forestry Commission Technical Paper 24.

Dreistadt S H, Clark J K, and Flint M L (1994). *Pests of Landscape Trees and Shrubs: An Integrated Pest Management Guide*, University of California Division of Agriculture and Natural Resources, Oakland, Publication 3359.

Fagerström T, Larsson S, Lohm U, and Tenow O (1978). Growth in Scots pine *Pinus sylvestris* L: a hypothesis on response to *Blastophagus piniperda* L Col Scolytidae attacks. *Forest Ecology and Management*, **1**: 273–281.

Fallon D J, Solter L F, Keena M, McManus M, Cate J R, and Hanks L M (2004). Susceptibility of Asian longhorned beetle, *Anoplophora glabripennis* (Motchulsky) (Coleoptera: Cerambyidae) to entomopathogenic nemtaodes. *Biological Control*, 30: 430–438.

Forestry Commission (2002). Dendroctonus micans – *a guide for Forest Managers in control techniques*. Plant Health Leaflet 9, HMSO, London.

Forestry Commission (2003). *Forestry Commission Consultation Paper on Controls Against the Spread of* Dendroctonus micans *the Great Spruce Bark Beetle*. http://www.forestry.gov.uk/website/pdf.nsf/pdf/consultationpaper1.pdf/$FILE/consultationpaper1.pdf

Frank J H and Foltz J L (1997). *Classical Biological Control of Pest Insects of Trees in the Southern United States: A Review and Recommendations*. US Department of Agriculture Forest Service, Forest Health Technology Enterprise Team, Technology Transfer Biological Control, Morgantown, WV, FHTET-96-20, p. 78.

Geiger C A and Gutierrez A P (2000). Ecology of *Heteropsylla cubana* Homoptera Psyllidae Psyllid Damage Tree Phenology Thermal Relations and Parasitism in the Field. *Environmental Entomology*, **29**(1): 76–86.

Global Invasive Species Database 2005. *Agrilus planipennis* available from: http://www.issg.org/database/species/ecology.asp.

Gregoire J-C (2003). *Spatial Ecology and Biological Control Research Group Website* http://lubies.ulb.ac.be/view.en.asp?thedoc=JCGEN

Griffiths M W (2001). The biology and ecology of *Hypsipyla* shoot borers. In: Floyd R B and Hauxwell C (eds), *Hypsipyla* Shoot Borers in Meliaceae, *Proceedings of an International Workshop*, Kandy, Sri Lanka, 20–23 August 1996, ACIAR Proceedings, No. 97, pp. 74–80.

Haack R A, Law K R., Mastro V C, Ossenbruggen H S, and Raimo B J (1997a). New York's battle with the Asian long-horned beetle. *Journal of Forestry*, **95**(12): 11–15.

Haack R A, Lawrence R K, Mccullough D G, and Sadof C S (1997b). *Tomicus piniperda* in North America an integrated response to a new exotic scolytid. In: Grégoire J C, Liebhold A M, Stephen M, Day K R, and Salom S (eds), *Proceedings Integrating Cultural Tactics into the Management of Bark Beetle and Reforestation Pests*, USDA Forest Service General Technical Report NE-236, pp. 62–72.

Haack R A, Jendek E, Liu H, Marchant K R, Petrice T R, Poland T M, and Ye H (2002). *The Emerald Ash Borer: A New Exotic Pest in North America. Newsletter of the Michigan Entomological Society*, **47**(3 and 4): 1–5.

Hagen, B (1999). New pests threaten urban eucalyptus. *Tree Notes*, Number 24, November, 5 pp.

Hanks L M, Gould J R, Paine T D, and Millar J G (1995). Biology and host relations of *Aventianella longoi* Siscaro (Hymenoptera: Encyrtidae), an egg parasitoid of the eucalyptus longhorned borer (Coleoptera: Cerambycidae). *Annals of the Entomological Society of America*, **88**: 666–671.

Hauxwell C (2001). Discussion summary silvicultural management of *Hypsipyla* spp. In: Floyd R B and Hauxwell C (eds), *Hypsipyla* Shoot Borers in Meliaceae, *Proceedings of an International Workshop*, Kandy, Sri Lanka, 20–23 August 1996, ACIAR Proceedings No. 97, p. 166.

Heritage S and Moore R (2001). *The Assessment of Site Characteristics as Part of a Management Strategy to Reduce Damage by* Hylobius. Forestry Commission Information Note. http://www.forestry.gov.uk/website/pdf.nsf/pdf/fcin38.pdf/$FILE/fcin38.pdf

Hertel G (1998). *Leucaena psyllid* Heteropsylla cubana. The Entomology and Forest Resources Digital Information Work Group, College of Agricultural and Environmental Sciences and Warnell School of Forest Resources, The University of Georgia, Tifton, Georgia 31793, USA, BUGWOOD 98-201. Originally produced as TCP/RAF/4451 A by the Food and Agriculture Organisation of the United Nations, Rome, Italy. Printed by Forest Health Management Centre, PO Box 30241, Nairobi.

Heydon D and Affonso M (1991). *Economic review of psyllid damage on* Leucaena *in Southeast Asia and Australia*. A report prepared for the Australian International Development Assistance Bureau, CAB, Wallingford, Oxon.

Horak M (2001). Current status of the taxonomy of *Hypsipyla Ragonot* Pyralidae Phycitinae. In: Floyd R B and Hauxwell C (eds), *Hypsipyla* Shoot Borers in Meliaceae, *Proceedings of an International Workshop*, Kandy, Sri Lanka, 20–23 August 1996, ACIAR Proceedings No. 97, pp. 69–73.

Humble L M and Stewart A J (1994). *Gypsy Moth*, Natural Resources Canada, Canadian Forest Service Pacific Forestry Centre, Victoria, BC, Forest Pest Leaflet 75, co-published by the BC Ministry of Forests. http://creatures.ifas.ufl.edu/urban/termites/formosan_termite.htm

King C J and Fielding N J (1989). Dendroctonus micans *in Britain – its Biology and Control*. Forestry Commission Bulletin 85, HMSO, London.

Koot H P (1991). *Spruce Aphid*, Forestry Canada Forest Insect and Disease Survey, Forest Pest Leaflet No. 16, 4 pp.

Lamb A B, Salom S M, and Kok L T (2005). Survival and reproduction of *Laricobius nigrinus* Fender (Coleoptera: Derodontidae), a predator of hemlock woolly adelgid, *Adelges tsugae* Annand (Homopetra: Adelgidae) in field cages. *Biological Control*, **32**: 200–207.

Larsson S and Tenow O (1980). Needle-eating insects and grazing dynamics in a mature Scots pine forest in central Sweden. In: Persson T (ed.), Structure and function of northern coniferous forests – an ecosystem study. *Ecological Bulletin*, **32**: 269–306.

Lowman M D (1984). An assessment of techniques for measuring herbivory: is rainforest defoliation more intense than we thought? *Biotropica*, **16**: 264–268.

McClure M S (1990). Role of wind, birds, deer, and humans in the dispersal of hemlock woolly adelgid (Homoptera: Adelgidae). *Environmental Entomology*, **19**: 36–43.

McClure M S (1996). Natural enemies of adelgids in North America: their prospect for biological control of *Adelges tsugae* (Homoptera: Adelgidae). In: Salom S M, Tigner T C, and Reardon R C (eds), *Proceedings of the First Hemlock Woolly Adelgid Review*, USDA Forest Service FHTET 96-10, pp. 89–101.

McClure M S (2005). *Hemlock Woolly Adelgid, Adelges tsugae (Annand)*, Connecticut Agricultural Experiment Station Fact Sheet EN012 (6/98). www.cases.state.ct.us/FactSheetFiles/Enotmology/fsen012fhtm

McCullough D G and Katovich S A (2004). *Pest Alert: Emerald Ash Borer*, USDA Forest Service Publication No. NA-PR-02-04.

MacFarlane D W and Meyer S P (2005). Characteristics and distribution of potential ash tree hosts for emerald ash borer. *Forest Ecology and Management*, **213**: 15–24.

MacLeod A, Evans H F, and Baker R H A (2002). An analysis of pest risk from an Asian longhorn beetle (*Anoplophora glabripennis*) to hardwood trees in the European community. *Crop Protection*, **21**: 635–645.

Mangoendihardjo S and Wagiman F X (1989). Some experiences in mass breeding release and evaluating the establishment of *Curinus coeruleus* Mulsant in Yogyakarta and Central Java. In: Napompeth B and MacDicken K (eds), *Proceedings Leucaena Psyllid Problems and Management*, Bogor, Indonesia, 16–21 January 1989. Winrock International, Bangkok, Thailand.

Millar J G, Paine T D, Campbell C D, and Hanks L M (2002). Methods for rearing *Syngaster lepidus* and *Jarra phoracantha* (Hymenoptera: Braconidae), larval parasitoids of the phloem-colonizing longhorned beetles *Phoracantha semipunctata* and *P. recurva* (Coleoptera: Cerambycidae). *Bulletin of Entomological Research*, **92**: 141–146.

Ministry of Forestry Canada Forest Practices Branch. www.for.gov.bc.ca/hfp/forsite/pest_field_guide/Green-spruce_aphid.htm

Morriscy S L (1996). *A Life-Table Study of the Effect of Parasitoids on the Bark Beetle* Ips typographus *L. Coleoptera Scolytidae*. MSc Dissertation, University of Wales, Bangor, 153 pp.

Nakahara L W, Nagamine S, Matayoshi B, Kumashiro, and Nakahara L M (1987). Biological control program of the leucaena psyllid *Heteropsylla cubana* Crawford Homoptera Psyllidae in Hawaii. In: Withington D and Brewbaker J L (eds), *Proceedings Workshop on Biological and Genetic Control Strategies for the Leucaena Psyllid*, 3–7 November 1986, Molokai and Honolulu, Hawaii. *Leucaena Research Report*, 7(2): 39–44.

Nomura S (2002). *Agrilus planipennis*. Canadian Food Inspection Agency Science Branch.

Oka I N (1989). Progress and future activities of the leucaena psyllid research program in Indonesia. In: Napompeth B and MacDicken K G (eds), *Proceedings Leucaena Psyllid Problems and Management*, Bogor, Indonesia, 16–21 January 1989. Winrock International, Bangkok, Thailand.

O'Neil C (1998). *Cypress Aphid Cinara cupressi*. The Entomology and Forest Resources Digital Information Work Group, BUGWOOD 98-202. www.Easternarc.org.html/98-202 html

Parry D, Spence J, and Volney W (1997). The response of natural enemies to experimentally increased populations of the forest tent caterpillar *Malacosoma disstria*. *Ecological Entomology*, **22**: 97–108.

Peters B C and Fitzgerald C J (2005). *Subterranean-Termite-Baiting systems*. DPI&F note. www.dpi.qld.gov.au/forestry/5040.html

Peters B C, King J, and Wylie F R (2005). *Subterranean Termites in Queensland*. DPI&F note. www.dpi.qld.gov.au/forestry/4974.html

Rose D and Leather S R (2003). http //www.bio.ic.ac.uk/research/srl/drose.htm

Sands D P A and Hauxwell C (2001). Discussion Summary Biological Control of *Hypsipyla* spp.

In: Floyd R B and Hauxwell C (eds), *Hypsipyla Shoot Borers in Meliaceae*, *Proceedings of an International Workshop*, Kandy, Sri Lanka, 20–23 August 1996, ACIAR Proceedings No. 97, pp. 146–147.

Scheffrahn R H and Su N-Y (1994). Keys to soldier and winged adult termites (Isoptera) of Florida. *Florida Entomologist*, 77: 460–475.

Smith M T, Zhong-qi Y, Hérard F, Fuester R, Bauer L, Solter L, Keena K, and D'Amico V (2005). Biological control of *Anoplophora glabripennis* (Motsch.): a synthesis of current research programs. http://www.uvm.edu/albeetle/research/biocontrol.html

Smitley D and McCullough D (2004). *How Homeowners Can Protect Ash Trees From the Emerald Ash Borer in Southeastern Michigan*. www.emeraldashborer.info

Sparks T C, Thompson G D, Kirst K A, Hertlein M B, Mynderse J S, Turner J R, and Worden T V (1998). Fermentation-derived insect control agents the spinosyns. In: Hall F R and Menn J J (eds), *Methods in Biotechnology Biopesticides Use and Delivery 5*, Humana Press, Totowa, NJ, pp. 171–188.

Speight M R and Wainhouse D (1989). *Ecology and Management of Forest Insects*, Oxford Scientific Publications, Oxford.

Speight M R, Hunter M D, and Watt A D (1999). *Ecology of Insects: Concepts and Applications*, Blackwell Science, Oxford.

Su N-Y and Scheffrahn R H (1987). Current status of the Formosan subterranean termite in Florida. In: Tamashiro M and Su N-Y (eds), *Biology and Control of the Formosan Subterranean Termite*, College of Tropical Agriculture and Human Resources, University of Hawaii, Honolulu, HI, pp. 27–31.

Tamashiro M and Su N-Y (eds) (1987). *Biology and Control of the Formosan Subterranean Termite*, College of Tropical Agriculture and Human Resources, University of Hawaii, Honolulu, HI.

Unger L S (1995). *Spruce Budworms*. Forest Pest Leaflet, Canadian Forest Service.

USDA (2002). *Asian Longhorned Beetle (Anoplophora glabripennis): A New Introduction*. Pest Alert NA-PR-01-99GEN.

Wagiman F X, Mangoendihardjo S, and Maahrub E (1989). Performance of *Curinus coeruleus* Mulsant as a predator against leucaena psyllid. In: Napompeth B and MacDicken K G (eds), *Proceedings Leucaena Psyllid Problems and Managment*, Bogor, Indonesia, 16–21 January 1989. Winrock International, Bangkok, Thailand, pp. 163–165.

Watson G W, Voegtlin D J, Murphy S T, and Foottit R G (1999). Biogeography of the *Cinara cupressi* complex Hemiptera Aphididae on Cupressaceae with description of a pest species introduced into Africa. *Bulletin of Entomological Research*, 89: 271–283.

Webber J and Eyre C (2003). A comparison of the *Ceratocystis* species associated with *Ips typographus* and *Ips cembrae*. Poster presentation, 8th *International Congress of Plant Pathology*, Christchurch, New Zealand. http://www.forestresearch.co.nz/PDF/11.31Webber&Eyre.pdf

Whittle K and Anderson D M (1985). *Spruce Bark Beetle* Ips typographus *L, Pests not Known to Occur in the United States or of Limited Distribution*. No. 66, USDA-APHIS-PPQ Ser 81-46, 13 pp.

Wingfield M J and Day R K (2002). *Risk of Exotic Forest Pests – Continental Overview for Sub-Saharan Africa*. Exotic pests online symposium. http://www.apsnet.org/online/exoticpest/Papers/wingfield.htm

Wood C S (1992). *Forest Tent Caterpillar*. Forestry Canada Forest Insect and Disease Survey Forest Pest Leaflet No. 17, 4 pp.

Wright M S, Osbrink W L, and Lax A R (2004). Potential of entomopathogenic fungi as biological control agents against the Formosan subterranean termite. In: Nelson W M (ed), *Agricultural Applications in Green Chemistry*, American Chemical Society Press Washington, DC, pp. 173–188.

Wylie F R (2001). Control of *Hypsipyla* spp. Shoot Borers with Chemical Pesticides: a Review. In: Floyd R B and Hauxwell C (eds), *Hypsipyla* Shoot Borers in Meliaceae. *Proceedings of an International Workshop*, Kandy, Sri Lanka, 20–23 August 1996, ACIAR Proceedings No. 97, pp. 109–115.

Yang X, Zhou J, Wang F, Cui M (1995). A study on the feeding habits of the larvae of two species of longicorn (*Anoplophora*) to different tree species. *Journal of Northwestern Forestry College*, **10**(2), 1–6 (in Chinese).

Yang Z and Smith M T (2001). Investigations of natural enemies for biocontrol of *Anoplophora glabripennis* (Motsch.). In: Fosbroke S L C and Gottschalk K W (eds), *Proceedings US Department of Agriculture Interagency Research Forum on Gypsy Moth and other Invasive Species*, Annapolis, MD, pp. 139–141.

Yorks T E, Jenkins J C, Leopold D J, Raynal D J, and Orwig D A (1999). Influences of eastern hemlock mortality on nutrient cycling. In: McManus K A, Shields K S, and Souto D R (eds), *Proceedings of the Symposium on Sustainable Management of Hemlock Ecosystems in Eastern North America*, USDA Forest Service General Technical Report NE-267, pp. 126–133.

Young R F, Shields K S, and Berlyn G P (1995). Hemlock woolly adelgid (Homoptera: Adelgidae): stylet bundle insertion and feeding sites. *Annals of the Entomological Society of America*, **88**: 827–835.

Zilahi-Balogh G M G, Kok L T and Salom S M (2002). Host specificity tests of *Laricobius nigrinus* Fender (Coleoptera: Derodontidae) a biological control agent of the hemlock woolly adelgid, *Adelges tsugae* (Homoptera: Adelgidae). *Biological Control*, **24**: 192–198.

CHAPTER 10 GENERAL FOREST ECOLOGICAL PROCESSES

REFERENCES

Ågren G I, Bosatta E, and Magill A H (2001). Combining theory and experiment to understand effects of inorganic nitrogen on litter decomposition. *Oecologia*, **128**: 94–98.

Anderson J M and Swift M J (1983). Decomposition in tropical forests. In: Sutton S L, Whitmore T C, and Chadwick A C (eds), *Tropical Rain Forest: Ecology and Management*, Blackwell Scientific Publications, Oxford, pp. 287–309.

Anderson W B and Eickmeier W G (1998). Physiological and morphological responses to shade and nutrient additions of *Claytonia virginica* (Portulacaceae): implications for the 'vernal dam' hypothesis. *Canadian Journal of Botany*, **76**:1340–1349.

Anderson W B and Eickmeier W G (2000). Nutrient resorption in *Claytonia virginica* L.: implications for deciduous forest nutrient cycling. *Canadian Journal of Botany*, **78**: 832–839.

Attiwill P M and Adams M A (1993). Nutrient cycling in forests. *New Phytologist*, **124**: 561–582.

Avissar R and Werth D (2005). Global hydroclimatological teleconnections resulting from tropical deforestation. *Journal of Hydrometeorology*, **6**: 134–145.

Barberis I M and Tanner E V J (2005). Gaps and root trenching increase tree seedling growth in Panamanian semi-evergreen forest. *Ecology*, **86**: 667–674.

Bates C G and Roeser J, Jr (1928). Light intensities required for growth of coniferous seedlings. *American Journal of Botany*, **15**: 185–194.

Beedlow P A, Tingey D T, Phillips D L, Hogsett W E, and Olszyk D M (2004). Rising atmospheric CO_2 and carbon sequestration in forests. *Frontiers in Ecology and the Environment*, **2**: 315–322.

Bobiec A (2002). Living stands and dead wood in the Białowieża forest: suggestions for restoration management. *Forest Ecology and Management*, **165**: 125–140.

Bosch J M and Hewlett J D (1982). A review of catchment experiments to determine the effect of vegetation changes on water yield and evapotranspiration. *Journal of Hydrology*, **55**: 3–23.

Brown A E, Zhang L, McMahon T A, Western A W, and Vertessy R A (2005). A review of paired catchment studies for determining changes in water yield resulting from alterations in vegetation. *Journal of Hydrology*, **310**: 28–61.

Calder I R, Reid I, Nisbet T R, and Green J C (2003). Impact of lowland forests in England on water resources: application of the Hydrological Land Use Change (HYLUC) model. *Water Resources Research*, **39**: paper 1319.

Domec J-C, Warren J M, Meinzer F C, Brooks J R, and Coulombe R (2004). Native root xylem embolism and stomatal closure in stands of Douglas-fir and Ponderosa pine: mitigation by hydraulic redistribution. *Oecologia*, **141**:7–16.

Emerman S H and Dawson T E (1996). Hydraulic lift and its influence on the water content of the rhizosphere: an example from sugar maple, *Acer saccharum. Oecologia*, **108**: 273–278.

Evans G C (1956). An area survey method of investigating the distribution of light intensity in woodlands, with particular reference to sunflecks, including an analysis of data from rain forest in Southern Nigeria. *Journal of Ecology*, **44**: 391–428.

FAO (2003). *State of the World's Forests 2003*, Food and Agriculture Organization of the United Nations, Rome, Italy.

Filella I and Penuelas J (2003–2004). Indications of hydraulic lift by *Pinus halepensis* and its effects on the water relations of neighbour shrubs. *Biologia Plantarum*, **47**: 209–214.

Fitter A H (2005). Darkness visible: reflections on underground ecology. *Journal of Ecology*, **93**: 231–243.

Franklin J F, Cromack K, Jr, McKee A, Masser C, Sedell J, Swanson F, and Juday G (1981). *Ecological Characteristics of Old-growth Douglas-fir Forests*. United States Department of Agriculture, Forest Service, Pacific Northwest Forest and Range Experiment Station, Portland, Oregon, General Technical Report PNW-118.

Golding D L (1970). The effects of forests on precipitation. *Forestry Chronicle*, **46**: 397–402.

Jackson R B, Canadell J, Ehleringer J R, Mooney H A, Sala O E, and Schulze E D (1996). A global analysis of root distributions for terrestrial biomes. *Oecologia*, **108**: 389–411.

Keppler F, Hamilton J T G, Braß M, and Röckmann T (2006). Methane emissions from terrestrial plants under aerobic conditions. *Nature*, **439**: 187–191.

Kirby K and Drake M (1993). Dead wood matters. *English Nature Science No 7*, English Nature, Peterborough.

Kuuluvainen T and Juntunen P (1998). Seedling establishment in relation to microhabitat variation in a windthrow gap in a boreal *Pinus sylvestris* forest. *Journal of Vegetation Science*, **9**: 551–562.

Likens G E (2004). Some perspectives on long-term biogeochemical research from the Hubbard Brook Ecosystem Study. *Ecology*, **85**: 2355–2362.

Likens G E, Driscoll C T, Buso D C, Siccama T G, Johnson C E, Lovett G M, Fahey T J, Reiners W A, Ryan D F, Martin C W, and Bailey S W (1998). The biogeochemistry of calcium at Hubbard Brook. *Biogeochemistry*, **41**: 89–173.

Lipson D and Näsholm T (2001). The unexpected versatility of plants: organic nitrogen use and availability in terrestrial ecosystems. *Oecologia*, **128**: 305–316.

Magill A H, Aber D A, Currie W S, Nadelhoffer K J, Martin M E, McDowell W H, Melillo J M and Steudler P (2004). Ecosystem response to 15 years of chronic nitrogen additions at the Harvard Forest LTER, Massachusetts, USA. *Forest Ecology and Management*, **196**:7–28.

Muller R N and Bormann F H (1976). Role of *Erythronium americanum* Ker. in energy flow and nutrient dynamics of a northern hardwood forest ecosystem. *Science*, **193**: 1126–1128.

Narukawa Y and Yamamoto S-I (2001). Gap formation, microsite variation and the conifer seedling occurrence in a subalpine old-growth forest, central Japan. *Ecological Research*, **16**:617–625.

Nosengo N (2003). Fertilized to death. *Nature*, **425**: 894–895.

Orr S P, Rudgers J A, and Clay K (2005). Invasive plants can inhibit native tree seedlings: testing potential allelopathic mechanisms. *Plant Ecology*, **181**: 153–165.

Penuelas J and Filella I (2003). Deuterium labelling of roots provides evidence of deep water access and hydraulic lift by *Pinus nigra* in a Mediterranean forest of NE Spain. *Environmental and Experimental Botany*, **49**: 201–208.

Poorter L, Zuidema P A, Pena-Claros M, and Boot R G A (2005). A monocarpic tree species in a polycarpic world: how can *Tachigali vasqueszii* maintain itself in a tropical rain forest. *Journal of Ecology*, **93**: 268–278.

Rothstein D E (2000). Spring ephemeral herbs and nitrogen cycling in a northern hardwood forest: an experimental test of the vernal dam hypothesis. *Oecologia*, **124**: 446–453.

Tessier J T and Raynal D J (2003). Vernal nitrogen and phosphorus retention by forest understorey vegetation and soil microbes. *Plant and Soil*, **256**: 443–453.

Thomas P (2000). *Trees: Their Natural History*, Cambridge University Press, Cambridge.

Thomas P A and Packham J R (2007). *Ecology of Woodlands and Forests*, Cambridge University Press, Cambridge.

Zak D R, Groffman P M, Pregitzer K S, Christensen S, and Tiedje J M (1990). The vernal dam: plant-microbe competition for nitrogen in northern hardwood forests. *Ecology*, **71**: 651–656.

CHAPTER 11 SILVICULTURAL SYSTEMS

REFERENCES

Hendrison J (1990). *Damage-controlled Logging in Managed Tropical Rainforest in Suriname*, Wageningen, The Netherlands.

ITTO (2004). *Annual Review and Assessment of the World Timber Situation*, Document GI-7/04, International Tropical Timber Organization, Yokohama, Japan.

Jonkers W B J (1987). *Vegetation Structure, Logging Damage and Silviculture in a Tropical Rainforest in Suriname*, Wageningen, The Netherlands.

Matthews J D (1989). *Silvicultural Systems*, Clarendon Press, Oxford.

Mergan F and Vincent R (eds) (1987). *Natural Management of Tropical Moist Forests: Silvicultural and Management Prospects of Sustained Utilization*, Yale University, Newhaven.

Palmer J and Synnott T J (1992). The management of natural forests. In Sharma N P (ed.), *Managing the World's Forests: Looking for a Balance Between Conservation and Development*, Kendall/Hunt Publishing Co., Iowa, pp. 337–373.

Parren M P E (1991). *Silviculture with Natural Regeneration: A Comparison Between Ghana, Cote d'Ivoire and Liberia*, AV no. 90/50, Department of Forestry, Wageningen Agricultural University, The Netherlands, p. 82.

Whitmore T C (1990). *An Introduction to Tropical Rainforest*, Clarendon Press, Oxford.

Wyatt-Smith J (1963). *Manual of Malayan Silviculture for Inland Forests*, Malayan Forestry Record No 23, Forestry Department, Malaya.

Wyatt-Smith J (1987). *The Management of Tropical Moist Forest for the Sustained Production of Timber: Some Issues*, IUCN/IIED Tropical Forestry Policy Paper, No. 4, p. 20.

CHAPTER 12 TREE PRUNING AND SURGERY IN ARBORICULTURE

REFERENCES

Shigo A L (1984).Compartmentalization: A conceptual framework for understanding how trees grow and defend themselves. *Annu. Rev. Phytopathol.*, **22**: 189–214.

Thomas P (2000). *Trees: Their Natural History*, Cambridge University Press, Cambridge.

GENERAL READING

Anonymous (1989). *Recommendations for Tree Work*, British Standards Institution BS3998.

Anonymous (2005). *A Guide to Good Climbing Practice*, Arboricultural Association/Forestry Commission.

Capel J A and Thorman D H (1994). *A Guide to Tree Pruning*, Arboricultural Association, Romsey, Hampshire.

Kirkham T (2004). *The Pruning of Trees and Shrubs*, 2nd edn, Timber Press, Portland, Oregon.

Lonsdale D (1999). *Principles of Tree Hazard Assessment and Management*, HMSO, London.

Mattheck C and Breloer H (1994). *The Body Language of Trees: A Handbook of Failure Analysis*. Research for Amenity Trees, no. 4, HMSO, London.

Read H (2000). *Veteran Trees: A Guide to Good Management*, English Nature.

Shigo A L (1991). *Modern Arboriculture*, Shigo and Tree Associates, Durham, New Hampshire.

CHAPTER 13 TREE PROPAGATION FOR FORESTRY AND ARBORICULTURE

REFERENCES

Brown N A C (1999). The role of fire in enhancing regeneration: the Cape floral region. In Bowes B G (ed.), *A Colour Atlas of Plant Propagation and Conservation*, Manson Publishing, London, pp. 157–167.

Hartmann H T, Kester D F, Davies F T, and Geneve R L (2002). *Plant Propagation. Principles and Practices*, Prentice Hall, Upper Saddle River, NJ.

Iriondo J M and Perez C (1999). Propagation from seeds and seed preservation. In Bowes B G (ed.), *A Colour Atlas of Plant Propagation and Conservation*, Manson Publishing, London, pp. 46–57.

Macdonald B (1986). *Practical Woody Plants Propagation for Nursery Growers, Volume 1*, Timber Press, Portland, OR.

Merkle S A (1999). Application of in vitro culture (ivc) for conservation of forest trees. In Bowes B G (ed.), *A Colour Atlas of Plant Propagation and Conservation*, Manson Publishing, London, pp. 119–130.

CHAPTER 14 FOREST AND WOODLAND CONSERVATION
REFERENCES

Baleé W (1987). *A etnobotânica quantitativa dos Índios Tembé (Rio Gurupi, Pará). Bol. Mus. Par. Emílio Goeldi Sér. Bot.*, 3(1): 29–50.

Batisse M (1986). Developing and focusing the biosphere reserve concept. *Nature and Resources*, 22: 1–10.

Boom B M (1987). Ethnobotany of the Chácobo Indians, Beni, Bolivia. *Advances in Economic Botany 5*, The New York Botanical Garden, NY, 68 pp.

Brook B W, Sodhi N S, and Ng P K L (2003). Catastrophic extinctions follow deforestation in Singapore. *Nature*, 424: 420–423.

Campbell D G, Daly D C, Prance G T, and Maciel U N (1986). Quantitative ecological inventory of terra firme and várzea tropical forest on the Rio Xingu, Brazilian Amazon. *Brittonia*, 38: 369–393.

Cox G W (1993). *Conservation Ecology*. W C Brown, Dubugue, Indiana.

Dransfield J (1989). The conservation status of rattans in 1987. A cause for great concern. In: Rao A N and Vongkaluang I (eds), *Recent Research on Rattans. Proceedings of the International Seminar*, 12–14 November 1987, Chiangmai, Thailand, Faculty of Forestry, Kasetsart University and IDRC Canada, pp. 6–10.

Dudley N and Stolton S (1999). *Threats to Forest Protected Areas*. Research Report from IUCN and the World Bank/WWF Alliance for Forest Conservation and Sustainable Use, IUCN, Gland, Switzerland.

Elias T S and Prance G T (1978). Nectaries on the fruit of *Crescentia* and other Bignoniaceae. *Brittonia*, 30: 175–181.

Frankel O H and Soulé M E (1981). *Conservation and Evolution*, Cambridge University Press, Cambridge.

Gadgil M (1983). Conservation of plant resources through biosphere reserves. In Jain S K and Mehra K L (eds), *Proceedings of the Regional Workshop on Conservation of Tropical Plant Resources in South-east Asia. Botanical Survey of India*, Howrah, India, pp. 66–71.

Gentry A H (1988). Tree species richness of upper Amazonian forests. *Proceedings of the US National Academy of Sciences*, 85: 156–159.

Godefroid S and Koedam N (2003). How important are large vs. small forest remnants for the conservation of the woodland flora in an urban context? *Global Ecology and Biogeography*, 12: 287–298.

Gunatilleke C V S, Gunatilleke I A U N, and Ashton P S (1995). Rainforest research and conservation; the Sinharaja experience in Sri Lanka. *The Sri Lanka Forester*, 22: 49–60.

Hawkes J G (1991). International workshop on dynamic *in situ* conservation of wild relatives of major cultivated plants: summary of final discussion and recommendations. *Israel Journal of Botany*, 40: 529–536.

Koski F T V (1991). *Preservation of Genetic Resources of Forest Trees in Finland*. Finnish Forest Research Institute, Vantaa, Finland (5p. Mimeo).

Lamb D, Parrotta R, Keenan R, and Tucker N (1997). Rejoining habitat fragments: restoring degraded rainforest lands. In Laurence W F and Bierregaard R O, Jr (eds), *Tropical Forest Remnants: Ecology, Management and Conservation of Fragmented Communities*. University of Chicago Press, Chicago, pp. 51–60.

Lawrence M J and Marshall D F (1997). Plant population genetics. In Maxted N, Ford-Lloyd B V, and Hawkes J G (eds), *Plant Genetic Conservation: The in situ Approach*, Chapman & Hall, London, pp. 99–131.

Lesica P and Allendorf F W (1992). Are small populations of plants worth preserving? *Conservation Biology*, 6: 135–139.

Lovejoy T E, Bierregaard R O, Jr, Rylands A B, Malcolm J R, Quintela C E, Harper L H, Brown K S, Jr, Powell A H, Powell G U N, Schubart H O R, and Hays M B (1986). Edge and other effects of isolation on Amazonian forest fragments. In Soulé M E (ed), *Conservation Biology: The Science of Scarcity and Diversity*, Sinauer Associates, Sunderland, Mass., pp. 257–285.

Murawski D A (1995). In Lowman M and Hamrick J L (eds), *Reproductive Biology and Genetics of Tropical Trees in Forest Canopies*, Academic Press, London.

Nepstad D and Schwartzman S (1992). Introduction to non-timber product extraction from tropical forests: Evaluation of a conservation and development strategy. *Advances in Economic Botany*, 9: vii–xii.

Newbury H J and Ford-Lloyd B V (1997). Estimation of genetic diversity. In Maxted N, Ford-Lloyd B V, and Hawkes J G (eds), *Plant Genetic Conservation*, Chapman & Hall, London, pp. 192–206.

O'Malley D M and Bawa K S (1987). Mating system of a tropical rain forest tree species. *American Journal of Botany*, 74: 1143–1149.

Peters C M and Hammond E J (1990). Fruits from the flooded forests of Peruvian Amazonia: yield estimates for natural populations of three promising species. *Advances in Economic Botany*, 8:159–176.

Powell A H and Powell G U N (1987). Population dynamics of male euglossine bees in Amazonian forest fragments. *Biotropica*, 19: 176–179.

Prance G T, Balée W, Boom B M, and Carneiro R L (1987). Quantitative ethnobotany and the case for conservation in Amazonia. *Conservation Biology*, 1: 296–310.

Pye-Smith C (2003). Fruits of the forest. *New Scientist*, 19 July: 36–39.

Rankin de Merona J M, Prance G T, Hutchings R W, da Silva M S, Rodrigues W A, and Vehling M E (1994). Preliminary results of large-scale tree inventory of upland rain forest in Central Amazon. *Acta Amazonica*, 22: 493–534.

Soulé M E (ed.) (1986). *Conservation Biology: The Science of Scarcity and Diversity*, Sinauer Associates, Sunderland, Mass.

Soulé M E (ed.) (1987). *Viable Populations for Conservation*, Cambridge University Press, Cambridge and New York.

Terborgh J (1986). Keystone plant resources in the tropical forest. In Soulé M E (ed.), *Conservation Biology: The Science of Scarcity and Diversity*, Sinauer Associates, Sunderland, Mass., pp. 330–334.

Terborgh J (1992). Maintenance of diversity in tropical forests. *Biotropica*, 24: 283–292.

Valencia R, Balsev H, and Paz y Mino G (1994). High tree alpha diversity in Amazonian Ecuador. *Biodiversity and Conservation*, 3: 21–28.

Vasquez R and Gentry A H (1989). Use and misuse of forest-harvested fruits in the Iquitos area. *Conservation Biology*, 3: 350–361.

Wilke D (1994). Carpe and non-wood forest products. In Sunderland T C H, Clark L E, and Vantomme P (eds), *Non-wood Forest Products of Central Africa: Current Research Issues and Prospects for Development*, FAO, Rome.

World Resources Institute (2001). *National and International Protection of Natural Areas*, World Resources Institute, Washington, DC, pp. 3–16.

Index